MEDICAL PHYSIOLOGY

TEXTBOOK STUDY GUIDE

By

RAUL N. DE GASPERI, M.D.

UNIVERSITY OF MIAMI
SCHOOL OF MEDICINE

*1225 Multiple Choice Questions
and Referenced Answers*

MEDICAL EXAMINATION PUBLISHING COMPANY, INC.
65-36 Fresh Meadow Lane
Flushing, New York 11365

Copyright © 1972 by the Medical Examination Publishing Co., Inc.

Copyright © 1972 by the Medical Examination Publishing Co., Inc.

All rights reserved. No part of this publication may be reproduced in any form or by any means, electronic or mechanical, including photocopy, without permission in writing from the publisher.

ISBN 0-87488-155-2

February 1972

PRINTED IN THE UNITED STATES OF AMERICA

PREFACE

Learning is a sequential process. The student is *exposed to a new concept* through books and lectures, and then *challenged* by problems to solve and questions to answer, using his newly acquired knowledge. Finally, his questions are *discussed* and answered, ideally, in a conference with his instructor.

Too often, however, a large student-teacher ratio precludes the possibility of discussion. In these instances, self-teaching is almost mandatory. For if the proper level of learning is to be attained, the learning process must remain constant in its three-part sequence, though new ways are substituted for old. These new ways, including a multiplicity of audio-visual aids and self-instructional programs, allow the student to reinforce the conceptual learning which is the first step, but not the final achievement, of the learning process.

This study guide should be used *after* a person has read a textbook of Physiology and attended a series of related lectures. The book selected to prepare this set of questions was *Textbook of Physiology* by Arthur C. Guyton, M.D., 4th edition, 1971. Questions were devised for the fundamental purpose of focusing on concepts which tend to remain invariant after a number of years. Intended as a contribution to self-teaching, this study guide provides the challenge referred to above as well as a ready source of answers to students' questions.

Although Physiology has the unique purpose of describing mechanisms and causal sequences, this description depends upon interpretation of experimental data. New interpretations are bound to appear, and thus, some concepts interpreted as correct now could change with time.

Should errors appear in the chosen answers, the fault is only that of the author of this study guide. It will be most gratefully acknowledged if these errors could be pointed out in order to avoid them in the future.

February, 1972
Coral Gables, Florida   Raul N. De Gásperi, M.D.

DEDICATION: To Graciela Finestra

TABLE OF CONTENTS

SECTION I	:	THE CARDIOVASCULAR SYSTEM Questions 1-268.	7
SECTION II	:	PHYSIOLOGY OF THE BODY FLUIDS AND THE KIDNEYS Questions 269-417.	56
SECTION III	:	PHYSIOLOGY OF THE RESPIRATORY SYSTEM Questions 418-545.	83
SECTION IV	:	PHYSIOLOGY OF THE NERVOUS SYSTEM Questions 546-912.	108
SECTION V	:	GASTROINTESTINAL PHYSIOLOGY Questions 913-977.	173
SECTION VI	:	METABOLISM, TEMPERATURE REGULATION AND HEPATIC PHYSIOLOGY Questions 978-1115	185
SECTION VII	:	ENDOCRINE PHYSIOLOGY Questions 1116-1225.	209
ANSWER KEY	:	. . . .	228

SECTION I - THE CARDIOVASCULAR SYSTEM

FOR EACH OF THE FOLLOWING MULTIPLE CHOICE QUESTIONS, SELECT THE ONE MOST APPROPRIATE ANSWER:

1. THE PROLONGED DEPOLARIZATION OF MYOCARDIAL FIBERS, CAUSING A RELATIVELY WIDE "PLATEAU" OF THE CARDIAC ACTION POTENTIALS, IS CORRELATED WITH:
   A. An increased potassium conductance
   B. A decreased sodium conductance
   C. A decreased potassium conductance
   D. A decreased permeability to calcium
   E. Increased activity of cholinesterase   Ref. p. 149

2. THE LENGTH OF THE FUNCTIONAL REFRACTORY PERIOD OF CARDIAC MUSCLE DETERMINES THE MINIMAL TIME THAT MUST ELAPSE BEFORE A SECOND CONTRACTION MAY OCCUR IN RESPONSE TO A SECOND ACTION POTENTIAL. NEURONS HAVE A SHORT REFRACTORY PERIOD AND SKELETAL MUSCLE FIBERS ARE REFRACTORY FOR A LONGER PERIOD THAN NEURONS AND AXONS. THE FUNCTIONAL REFRACTORY PERIOD OF VENTRICULAR MUSCLE FIBERS IS APPROXIMATELY:
   A. 10 milliseconds
   B. 50 milliseconds
   C. 250 milliseconds
   D. 1,000 milliseconds
   E. 2,000 milliseconds   Ref. p. 150

3. BY COMPARISON WITH THE VENTRICULAR MUSCLE REFRACTORY PERIOD, THE FUNCTIONAL REFRACTORY PERIOD OF THE ATRIAL MUSCLE IS:
   A. Shorter
   B. Longer
   C. The same   Ref. p. 150

4. THE REASON WHICH PROBABLY UNDERLIES THE CORRELATION BETWEEN VOLTAGE AND DURATION OF THE ACTION POTENTIAL OF MYOCARDIAL FIBERS ON ONE SIDE AND THE WORK OF CONTRACTION OF THESE FIBERS IS THAT WITH GREATER ELECTRICAL CHANGES THERE IS A:
   A. More rapid influx of sodium
   B. More rapid efflux of potassium
   C. Larger quantity of calcium released from the sarcoplasmic reticulum
   D. Greater inhibition of ATP-ase   Ref. p. 150

5. THE PERCENTAGE OF THE TOTAL TIME (CONTRACTION PLUS RELAXATION) REQUIRED FOR ONE HEART BEAT VARIES WITH THE RATE OF CONTRACTION. THIS IS IMPORTANT TO REMEMBER BECAUSE IT IS DURING RELAXATION THAT THE HEART IS ABLE TO FILL ITS CHAMBERS PREPARING FOR THE NEXT CONTRACTION AND EJECTION. DURING A TACHYCARDIA OF INCREASING SEVERITY, THE TIME "USED UP" BY THE CONTRACTION PHASE IN RELATION TO THE TOTAL TIME IS KNOWN TO BE:
   A. Increasingly larger
   B. Increasingly less
   C. Constant
   D. Variable and actually biphasic   Ref. p. 150

SECTION I - THE CARDIOVASCULAR SYSTEM

6. IN THE NORMAL INDIVIDUAL, BLOOD CONTAINED IN THE TWO ATRIA IS SUBJECTED TO PRESSURES WHICH VARY DURING THE CARDIAC CYCLE AND WHICH ARE ALSO DEPENDENT ON ATRIAL COMPLIANCE AND RATE OF ATRIAL FILLING. WHEN IN THE NORMAL INDIVIDUAL THE LEFT ATRIAL PRESSURE IS COMPARED WITH THAT IN THE RIGHT ATRIUM IT IS FOUND THAT LEFT ATRIAL PRESSURE IS:
   A. Greater than right atrial pressure at all times during one cardiac cycle
   B. Less than right atrial pressure at all times during one cardiac cycle
   C. Greater than right atrial pressure during atrial systole only
   D. Less than right atrial pressure during atrial systole only
   Ref. p. 152

7. THE CREST OF THE V WAVE IN ATRIAL PRESSURE RECORDINGS MARKS THE PRECISE TIME WHEN:
   A. Atrial contraction occurs
   B. Ventricular contraction occurs
   C. Aortic and pulmonary valves close at the end of systole
   D. Mitral and tricuspid valves open at the end of isovolumetric period
   E. The Q wave of the electrocardiogram is inscribed
   Ref. p. 152

8. THE TERM "PROTODIASTOLE" IS GIVEN TO THE PHASE OF THE CARDIAC CYCLE WHEN:
   A. Ventricles are closed cavities and relaxing isovolumetrically
   B. Ventricles are in communication with the aorta and pulmonary arteries but are no longer ejecting a significant quantity of blood
   C. The atrial diastole that coincides with maximal ventricular ejection
   D. Ventricles are filling rapidly with blood previously accumulated in the atria
   Ref. p. 153

9. APPROPRIATE AND OPTIMAL CLOSURE OF THE MITRAL AND TRICUSPID VALVES OCCURS WHEN PAPILLARY MUSCLES OF THE CORRESPONDING VALVE "APPARATUS":
   A. Relax and become elongated during systole
   B. Contract and become shortened during systole
   C. Are subjected to passive tension
   D. Do not pull on the chordae tendinae
   Ref. p. 153

10. THE RATE AT WHICH INTRAVENTRICULAR PRESSURE IS DEVELOPED DURING VENTRICULAR CONTRACTION CHANGES DURING VARIOUS PHASES OF VENTRICULAR SYSTOLE. ITS HIGHEST INSTANTANEOUS VALUE MAY BE FOUND:
    A. Before the aortic and pulmonary valves are open
    B. Immediately after the aortic and pulmonary valves are open, during the phase of ejection
    C. Just before the aortic and pulmonary valves close
    D. Just after closure of aortic and pulmonary valves
    Ref. p. 154

SECTION I - THE CARDIOVASCULAR SYSTEM

11. ALTHOUGH THE FORCE OF VENTRICULAR CONTRACTION CAN VARY AND THE EMPTYING OF ITS CAVITY MAY BE VARIABLE ACCORDING TO THE FORCE OF THE CONTRACTION, THE <u>END SYSTOLIC</u> VOLUME OF BLOOD LEFT WITHIN THE VENTRICLES OF A NORMAL ADULT HEART BEATING NORMALLY EQUALS ABOUT:
    A. 5 ml
    B. 50 ml
    C. 500 ml
    D. 700 ml
    E. Zero   Ref. p. 154

12. BLOOD CIRCULATES THROUGH VESSELS BECAUSE EACH UNIT VOLUME OF BLOOD ACQUIRES ENERGY AS IT PASSES THROUGH THE HEART. THE HEART TRANSFORMS CHEMICAL ENERGY INTO MECHANICAL ENERGY. THE BLOOD MOVING RAPIDLY IN THE AORTA HAS MOST OF ITS ENERGY IN THE FORM OF:
    A. Potential (pressure) energy
    B. Kinetic (velocity) energy
    C. Volumetric energy
    D. Tension energy   Ref. p. 155

13. THE AMOUNT OF WORK DONE BY THE HEART DURING A CERTAIN TIME IS PROPORTIONAL TO THE:
    A. Velocity and the volume of blood circulating through the heart
    B. Diastolic time
    C. Tension developed in the heart muscle and the time of contraction
    D. Reciprocal of the radius and the time of contraction
    E. Level of arterial blood pressure only
                                              Ref. p. 155

14. THE <u>EFFICIENCY</u> OF THE HEART INCREASES WHEN FOR A GIVEN VALUE OF CARDIAC WORK OUTPUT DURING A PERIOD OF TIME THE:
    A. Rate of oxygen consumption is relatively high
    B. Rate of oxygen consumption is relatively low
    C. Rate of carbon dioxide expired increases
    D. Rate of carbon dioxide expired decreases
    E. Respiratory exchange ratio is maintained constant
                                              Ref. p. 155

15. IF THE VENOUS RETURN TO THE RIGHT ATRIUM SHOULD INCREASE FROM A NORMAL VALUE OF ABOUT 5 LITERS/MINUTE (IN AN ADULT) TO ABOUT 10 LITERS/MINUTE, A NORMAL HEART WILL MOST LIKELY DEVELOP IN THE SUCCEEDING TWO MINUTES:
    A. An increase in atrial and end-diastolic right ventricular pressure with decreased contractility
    B. A decrease in atrial and end-diastolic right ventricular pressure with decreased contractility
    C. An increased atrial and end-diastolic right ventricular pressure with increased contractility
    D. Bradycardia and cardiac shock
    E. Bradycardia and very high end-diastolic pressure of the right ventricle   Ref. p. 156

SECTION I - THE CARDIOVASCULAR SYSTEM

16. FROM INSPECTION OF VENTRICULAR FUNCTION CURVES OBTAINED IN NORMAL DOGS (CAN BE EXTRAPOLATED TO HUMANS) FOR THE LEFT VENTRICLE AND FOR THE RIGHT VENTRICLE IT IS POSSIBLE TO CONCLUDE THAT FOR A SIMILAR VALUE OF INTRA-ATRIAL PRESSURE (I.E. 5 mm Hg) THE RATIO $\frac{\text{RIGHT VENTRICULAR WORK}}{\text{LEFT VENTRICULAR WORK}}$ IS ON THE ORDER OF:
    A. 2.0
    B. 1.0
    C. 0.1
    D. 0.5
    E. 1.5   Ref. p. 156

17. A DECREASED AVAILABILITY OF SYMPATHETIC NEUROTRANSMITTERS WILL PRODUCE CARDIAC EFFECTS WHICH WILL RESULT IN:
    A. Increased force of contraction with a slower rate of contraction
    B. Decreased force of contraction and a faster rate of contraction
    C. Decreased force of contraction and slower rate of contraction
    D. Increased force of contraction and faster rate of contraction
                                         Ref. p. 158

18. HYPERKALEMIA, AN ABNORMALLY HIGH CONCENTRATION OF POTASSIUM IONS (6-9 mEq/L) IN THE EXTRACELLULAR FLUID, MAY CAUSE DEATH BY:
    A. A decreased contractility of the heart muscle
    B. A higher contractility of the heart muscle which precipitates ventricular tachycardia
    C. No change in cardiac contractility but an irreversible increased resting potential of the myocardial sino-atrial fibers
    D. Stopping the heart in a pseudo-tetanic systole
                                         Ref. p. 159

19. THE HEART-LUNG PREPARATION OF STARLING HAS BEEN VERY USEFUL IN DEMONSTRATING THE RELATIONSHIP BETWEEN END-DIASTOLIC VOLUME AND STROKE OUTPUT. IN ADDITION IT HAS DEMONSTRATED THAT THE INCREASE IN "AORTIC" PRESSURE WITHIN THE LIMITS FOUND IN A MODERATE HYPERTENSION (MEAN PRESSURE UP TO 200 mm Hg) WILL PRODUCE IN A NORMAL HEART:
    A. A significant increase in systolic stroke volume
    B. No significant change in systolic stroke volume
    C. A significant decrease in systolic stroke volume
    D. An increase in heart rate
    E. A decrease in left ventricular volume
                                         Ref. p. 160

20. A LIKELY EXPLANATION FOR THE LOWER RESTING MEMBRANE POTENTIAL OF SINO-ATRIAL PACEMAKER FIBERS AND THEIR AUTOMATICITY IS THAT THESE SPECIALIZED FIBERS HAVE A(N):
    A. Increased permeability to potassium
    B. Decreased permeability to potassium
    C. Decreased permeability to sodium
    D. Increased permeability to sodium
    E. Decreased permeability to calcium   Ref. p. 162

SECTION I - THE CARDIOVASCULAR SYSTEM

21. THE RECOVERY OF THE ORIGINAL TRANSMEMBRANE CARDIAC POTENTIAL DURING THE EARLIEST PHASE OF ELECTRICAL DIASTOLE CAN BE BEST EXPLAINED BY A VERY SIGNIFICANT:
    A. Increase in potassium conductance
    B. Decrease in potassium conductance
    C. Increase in sodium conductance
    D. No change in sodium conductance
    E. Increase in calcium permeability   Ref. p. 162

22. SINO ATRIAL MYOCARDIAL PACEMAKER FIBERS ARE DISTINCT FROM MYOCARDIAL FIBERS IN THAT DURING THE TIME LAPSE BETWEEN REPETITIVE ACTION POTENTIALS THE SODIUM CONDUCTANCE SEEMS TO BE UNIFORMLY HIGH WHILE THE POTASSIUM CONDUCTANCE IS:
    A. Continuously increasing from a low level
    B. Continuously decreasing from a high level
    C. Undergoing no change at all
    D. Going through a biphasic change, high first, low later, and finally high again at the end of the electrical diastole
              Ref. p. 162

23. THE GREATEST DELAY IN CONDUCTION OF IMPULSES ORIGINATED IN THE SINOATRIAL NODE BEFORE THEY ENTER THE ATRIOVENTRICULAR BUNDLE TAKES PLACE IN THE:
    A. Cells of the A-V node
    B. Myocardial cells of the atrium
    C. Junctional fibers connecting the atrium with the A-V node
    D. Transitional fibers connecting the A-V node with the A-V bundle
              Ref. p. 163

24. ACETYLCHOLINE RELEASED BY THE TERMINAL ENDS OF PARASYMPATHETIC FIBERS IS KNOWN TO SLOW THE HEART BY AN EFFECT ON THE SINOATRIAL NODE AND ATRIOVENTRICULAR NODE FIBERS. IT ALSO HAS AN EFFECT ON THE ATRIAL MUSCLE WHICH HAS RELEVANCE IN TERMS OF ATRIAL FIBRILLATION. ATRIAL MUSCLE HAS BEEN SHOWN TO RESPOND TO ACETYLCHOLINE BY A(N):
    A. Increase in the magnitude of the atrial action potential
    B. Prolongation of the refractory period
    C. Shortening of the refractory period
    D. Increased time required for the upstroke of the atrial action potential
    E. Flattening of the whole action potential
              Ref. Fig. 14-4, p. 164

25. THE REPETITIVE TENDENCY OF PACEMAKER CELLS (SINOATRIAL NODE, ATRIOVENTRICULAR NODE) TO UNDERGO A GRADUAL DEPOLARIZATION REQUIRES A CERTAIN LENGTH OF TIME. THE TIME REQUIRED TO REACH THRESHOLD AND FIRE AN ACTION POTENTIAL IS SHORTER IN THE:
    A. Sinoatrial node fibers
    B. Atrioventricular node fibers
    C. Purkinje fibers   Ref. p. 165

SECTION I - THE CARDIOVASCULAR SYSTEM

26. THE BASIC MECHANISM THROUGH WHICH ACETYLCHOLINE, RELEASED FROM VAGUS NERVE TERMINALS, CAUSES BRADYCARDIA IS THROUGH THE INDUCTION OF A:
    A. Greater amount of calcium released from the sarcoplasm in pacemaker cells
    B. Greater sodium conductance in pacemaker cells
    C. Higher potassium permeability in pacemaker cells
    D. Decreased refractory period of pacemaker cells
    E. All of the above   Ref. p. 166

27. VENTRICULAR ESCAPE OCCURS DURING MILD VAGAL STIMULATION BECAUSE THE:
    A. Refractory period of the A-V nodal fibers is overcome by larger action potentials arriving from the S-A node
    B. A-V node is not influenced at all by the action of acetylcholine, and thus is able to take over the pacemaker activity suppressed in the S-A node
    C. Ventricles are stimulated by the sympathetics as a compensatory safeguarding mechanism
    D. Purkinje cells become pacemakers   Ref. p. 166

28. ALTHOUGH THE MECHANISM OF ACTION OF EPINEPHRINE ON CARDIAC MUSCLE FIBERS IS STILL A MATTER OF SPECULATION, A LOGICAL TENTATIVE EFFECT OF EPINEPHRINE IS THE INDUCTION OF A:
    A. Greater permeability for sodium ions
    B. Greater permeability for potassium ions
    C. Diminished permeability for sodium ions
    D. Diminished permeability for calcium ions
                                            Ref. p. 167

29. THE RAPID STIMULATION OF ATRIAL MUSCULATURE FOUND IN ATRIAL FLUTTER HAS BEEN EXPLAINED BY A WAVE OF DEPOLARIZATION WHICH PROGRESSES ALONG A CIRCULAR (CIRCUS MOVEMENT) OR ELLIPSOIDAL COURSE WHICH HAS A SELF-PERPETUATING TENDENCY. ATRIAL CONDITIONS WHICH SHOULD FAVOR THE CONTINUITY OF SUCH A DEPOLARIZING WAVE ARE:
    A. Shortened refractory period, dilated atria and decreased rate of propagation of the depolarizing wave (conduction)
    B. Prolonged refractory period, dilated atria and fast conduction
    C. Prolonged refractory period, dilated atria and slow conduction
    D. Shortened refractory period, dilated atria and fast conduction
                                            Ref. p. 168

30. A VERY USEFUL PROTECTIVE DEVICE "BUILT IN" THE HEART IS THE RELATIVELY PROLONGED (0.3 SECONDS) REFRACTORY PERIOD OF THE A-V NODAL FIBERS. THIS PREVENTS ANOTHER STIMULUS FROM REACHING THE VENTRICLES BEFORE THIS TIME IS OVER. ATRIAL FIBRILLATION IS FREQUENTLY ASSOCIATED WITH VENTRICULAR RATES IN THE RANGE OF:
    A. 35-67 irregular beats/minute
    B. 65-100 regular beats/minute
    C. 100-200 irregular beats/minute
    D. Over 300 regular beats/minute   Ref. p. 170

SECTION I - THE CARDIOVASCULAR SYSTEM

31. VENTRICULAR FIBRILLATION IS MORE EASILY CAUSED BY:
    A. Prolongation of refractory period, a uniform spatial distribution of the depolarizing wave, and infrequent ventricular stimuli
    B. Shortening of refractory period, an irregular spatial distribution of the depolarizing wave, and frequent ventricular stimuli
    C. Rapid atrial premature beats
    D. Sinoatrial tachycardia with intermittent A-V block
    Ref. p. 170

32. THE RELATIVELY LONG PERIOD OF TIME (200 MILLISECONDS) DURING WHICH THE VENTRICULAR CARDIAC MUSCLE MEMBRANE IS DEPOLARIZED (THE PLATEAU OF VENTRICULAR ACTION POTENTIALS) COINCIDES WITH A CERTAIN PORTION OF THE <u>PERIPHERALLY</u> OBTAINED ELECTROCARDIOGRAM. THIS IS THE:
    A. P wave
    B. P-Q interval
    C. Q wave
    D. S-T segment
    E. T wave   Ref. p. 174

THE NEXT THREE QUESTIONS REFER TO THE SAME TOPIC. THE VARIOUS CONVENTIONAL STANDARD ELECTROCARDIOGRAPHIC LIMB LEADS HAVE EMPIRICALLY BEEN OBTAINED BY MAKING CONNECTIONS BETWEEN THE NEGATIVE (-) AND POSITIVE(+) TERMINALS OF A SENSITIVE GALVANOMETER AND THREE EXTREMITIES: THE RIGHT ARM (RA), THE LEFT ARM (LA), AND THE LEFT LEG (LL). FIVE CONNECTIONS ARE LISTED BELOW, AND ITEMIZED BY CAPITAL LETTERS:

    A. LL(-)-------RA(+)
    B. LA(-)-------LL(+)
    C. LA(-)-------RA(+)
    D. RA(-)-------LL(+)
    E. RA(-)-------LA(+)

33. INDICATE WHICH OF THESE CONNECTIONS IS USED FOR LEAD I:
    A. ___   D. ___
    B. ___   E. ___
    C. ___   Ref. p. 177

34. INDICATE WHICH OF THESE CONNECTIONS IS USED FOR LEAD II:
    A. ___   D. ___
    B. ___   E. ___
    C. ___   Ref. p. 177

35. INDICATE WHICH OF THESE CONNECTIONS IS USED FOR LEAD III:
    A. ___   D. ___
    B. ___   E. ___
    C. ___   Ref. p. 177

SECTION I - THE CARDIOVASCULAR SYSTEM

THE NEXT FIVE QUESTIONS ALSO REFER TO PLACEMENT OF ELECTRODES TO RECORD AN ELECTROCARDIOGRAM. A 5,000 OHMS RESISTOR IS DESIGNATED BY THE ABBREVIATION (res.) AND IS USED BETWEEN THE EXTREMITY WHERE THE CONNECTION IS BEING PLACED AND THE NEGATIVE TERMINAL OF THE GALVANOMETER TO CONTRIBUTE TO THE CENTRAL TERMINAL OF WILSON. THE ABBREVIATIONS RA, res; LA, res.; LL, res. (-) MEANS THAT THE NEGATIVE TERMINAL OF THE GALVANOMETER HAS BEEN CONNECTED WITH THE RIGHT ARM (RA) THROUGH A RESISTOR (res.); THE LEFT ARM (LA) AND THE LEFT LEG (LL) WERE ALSO SIMILARLY CONNECTED THROUGH RESISTORS OF THE SAME MAGNITUDE TO THE SAME NEGATIVE TERMINAL OF THE GALVANOMETER. FIVE CONNECTIONS ARE LISTED BELOW, AND ITEMIZED BY CAPITAL LETTERS:

        CENTRAL TERMINAL   EXPLORING ELECTRODE

A. RA, res.; LA, res.; LL, res.(-) ---- Left Mid clavicular line at 5th i.c. space
B. RA, res.; LA, res.; LL, res.(-) ---- LL(+)
C. RA, res.; LA, res.; (-) ------------ LL(+)
D. RA, res.; LA, res.; LL, res. (-) ---- 4th intercostal space immediately to the right of the sternal border (+)
E. RA, res.; LA, res.; LL, res.(-) ---- 5th left i.c. space at left mid axillary line (+)

36. INDICATE WHICH OF THESE CONNECTIONS IS USED TO RECORD V-1:
    A. ___   D. ___
    B. ___   D. ___
    C. ___   Ref. p. 178

37. INDICATE WHICH OF THESE CONNECTIONS IS USED TO RECORD V-F:
    A. ___   D. ___
    B. ___   E. ___
    C. ___   Ref. p. 178

38. INDICATE WHICH OF THESE CONNECTIONS IS USED TO RECORD aV-F:
    A. ___   D. ___
    B. ___   E. ___
    C. ___   Ref. p. 178

39. INDICATE WHICH OF THESE CONNECTIONS IS USED TO RECORD V-4:
    A. ___   D. ___
    B. ___   E. ___
    C. ___   Ref. p. 178

40. INDICATE WHICH OF THESE CONNECTIONS IS USED TO RECORD V-6:
    A. ___   D. ___
    B. ___   E. ___
    C. ___   Ref. p. 178

SECTION I - THE CARDIOVASCULAR SYSTEM

THE FOLLOWING FIVE QUESTIONS ARE CONCERNED WITH VECTORS WHICH INDICATE THE DIRECTION IN WHICH ELECTRIC FIELDS MOVE AS DEPOLARIZATION WAVES IN THE HEART MUSCLE. A CLEAR PICTORIAL MEMORIZATION OF THE DIRECTION OF THE "AXIS" OF THE FRONTAL ELECTROCARDIOGRAPHIC LEADS IS USEFUL TO UNDERSTAND VECTORIAL ANALYSIS.

IN THE <u>FRONTAL</u> PLANE, THE ANGULAR ROTATION FORMING POSITIVE ANGLES HAS BEEN DEFINED IN ELECTROCARDIOGRAPHY AS RUNNING CLOCKWISE STARTING FROM A ZERO ANGLE DEFINED BY THE HORIZONTAL LINE RUNNING FROM THE RIGHT SHOULDER TO THE LEFT SHOULDER OR A SIMILAR HORIZONTAL LINE PARALLEL TO THIS, RUNNING THROUGH THE CENTER OF THE CHEST FROM RIGHT TO LEFT. ANGULAR ROTATION IN THE OPPOSITE DIRECTION (COUNTERCLOCKWISE) DRAWS NEGATIVE ANGLES.

THE ORIGIN OF THE LINE DRAWN BY THE VECTOR CORRESPONDS TO THE NEGATIVE SIDE OF THE DEPOLARIZATION WAVE; THE END TO THE POSITIVE SIDE.

A. Horizontal, right arm (-) to left arm (+) $0°$
B. Line connecting right shoulder (-) to pubis (+) $+60°$
C. Line connecting left shoulder (-) to pubis (+) $+120°$
D. Line connecting center of chest (-) vertically down to pubis (+) $+90°$
E. Line connecting center of chest (-) with right shoulder (+) $-150°$
F. Line connecting center of chest (-) with left shoulder (+) $-30°$
G. Horizontal, left arm (-) to right arm (+) $+180°$
H. Line connecting center of chest (-) vertically up to head (+) $-90°$

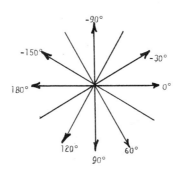

41. INDICATE WHICH LETTER IN THE PRECEDING LIST OF VECTORS DESCRIBES THE ANGLE DRAWN BY aVR:
    A. ___   E. ___
    B. ___   F. ___
    C. ___   G. ___
    D. ___   H. ___
    Ref. pp. 181-182

SECTION I - THE CARDIOVASCULAR SYSTEM

42. INDICATE WHICH LETTER DESCRIBES THE ANGLE DRAWN BY aVL:
    A. ___   E. ___
    B. ___   F. ___
    C. ___   G. ___
    D. ___   H. ___
                  Ref. pp. 181-182

43. INDICATE WHICH LETTER DESCRIBES THE ANGLE DRAWN BY LEAD II:
    A. ___   E. ___
    B. ___   F. ___
    C. ___   G. ___
    D. ___   H. ___
                  Ref. pp. 181-182

44. INDICATE WHICH LETTER DESCRIBES THE ANGLE DRAWN BY LEAD III:
    A. ___   E. ___
    B. ___   F. ___
    C. ___   G. ___
    D. ___   H. ___
                  Ref. pp. 181-182

45. INDICATE WHICH LETTER DESCRIBES THE ANGLE DRAWN BY LEAD aVF:
    A. ___   E. ___
    B. ___   F. ___
    C. ___   G. ___
    D. ___   H. ___
                  Ref. pp. 181-182

46. SUPPOSE THAT 0.04 SECONDS AFTER THE INITIATION OF VENTRICULAR DEPOLARIZATION, LEAD I RECORDS A POSITIVE DEFLECTION EQUAL TO 0.8 mV (8 mms), LEAD II A POSITIVE DEFLECTION EQUAL TO 0.8 mV (8 mms) AND LEAD III RECORDS ZERO mV (REMAINS ISOELECTRIC). ONE MAY CONCLUDE THAT DURING THIS TIME THE VECTOR POINTING TO THE DIRECTION OF DEPOLARIZATION IS:
    A. Halfway between leads I and II and perpendicular to lead III
    B. Parallel to lead II and perpendicular to aVL
    C. Parallel to aVF and perpendicular to lead I
    D. Parallel to lead II   Ref. pp. 182-183

47. IN THE CASE DESCRIBED BY THE PRECEDING QUESTION THE VECTOR DRAWS AN ANGLE WITH THE HORIZONTAL EQUAL TO:
    A. $+90°$
    B. $+0°$
    C. $+60°$
    D. $+30°$
    E. $+120°$ Ref. pp. 182-183

SECTION I - THE CARDIOVASCULAR SYSTEM

48. A Q WAVE RECORDED IN LEAD I IS BELIEVED TO BE DUE TO DEPOLARIZATION OF THE:
    A. Last portion of the A-V node
    B. Right bundle branch whose vector points toward the apex
    C. Left side of the base of heart at the end of ventricular depolarization
    D. Atria in a retrograde fashion
    E. Left side of the septum whose vector points toward the right shoulder
    Ref. p. 183

49. NORMAL REPOLARIZATION OF THE VENTRICLES USUALLY BEGINS IN THE AREA CLOSE TO THE:
    A. Lower part of the septum
    B. Endocardial surface of both ventricles
    C. Epicardial surface near the apex
    D. Upper part of the septum on the left side
    E. Base of the heart   Ref. p. 184

50. REPOLARIZATION OF THE ATRIAL MUSCLE:
    A. Occurs approximately simultaneously with the ventricular repolarization
    B. Coincides with the P-Q interval
    C. Is largely obliterated by the concomitant ventricular depolarization
    D. Has a vector which parallels the direction and has the same polarity as that of ventricular depolarization   Ref. p. 185

51. POSITIVE AND NEGATIVE DEFLECTIONS ARE ALGEBRAICALLY ADDED IN AT LEAST TWO FRONTAL LEADS IN ORDER TO CALCULATE THE DIRECTION OF THE MEAN ELECTRICAL AXIS OF VENTRICULAR DEPOLARIZATION. SUPPOSE THAT LEAD I INSCRIBES A POSITIVE DEFLECTION EQUAL TO 1.0 mV, AND A NEGATIVE DEFLECTION OF 0.25 mV; LEAD II INSCRIBES A POSITIVE DEFLECTION EQUAL TO 1.5 mV; LEAD III INSCRIBES A POSITIVE DEFLECTION EQUAL TO 1.0 mV AND A NEGATIVE DEFLECTION EQUAL TO 0.25 mV. THE DIRECTION OF THE MEAN ELECTRICAL AXIS IS:
    A. $0°$
    B. $45°$
    C. $60°$
    D. $90°$
    E. $120°$ Ref. p. 186

52. SUPPOSE THAT AN ELECTROCARDIOGRAM HAS QRS DEFLECTIONS WHICH SHOW THE FOLLOWING MAGNITUDES IN THE THREE STANDARD LEADS: LEAD I NEGATIVE 0.75; LEAD II POSITIVE 0.75; LEAD III POSITIVE 1.5. THE DIRECTION OF THE MEAN ELECTRICAL AXIS IS:
    A. $0°$
    B. $45°$
    C. $60°$
    D. $90°$
    E. $120°$ Ref. p. 186

SECTION I - THE CARDIOVASCULAR SYSTEM

53. IF AN ELECTROCARDIOGRAM OF A 60 YEAR-OLD MAN SHOWS THAT THE MEAN ELECTRICAL AXIS OF THE QRS COMPLEX IS POINTED ALONG THE -30° AXIS, THIS FINDING MAY BE INTERPRETED AS COMPATIBLE WITH, BUT NOT DIAGNOSTIC OF, A:
    A. Right ventricular hypertrophy
    B. Left ventricular hypertrophy
    C. Left bundle block
    D. Vertical heart in a slender person
    E. Normal deviation   Ref. p. 187

54. AN ELECTROCARDIOGRAM WHERE QRS COMPLEXES LAST 0.16 SECONDS MOST LIKELY DEMONSTRATES A:
    A. Partial block at the level of the A-V node
    B. Transmural infarction
    C. Ventricular hypertrophy
    D. Normal rhythm
    E. Complete bundle branch block   Ref. p. 190

55. IF THE LEVEL OF THE ZERO POTENTIAL LINE (ZERO p.l.) IS DEFINED BY THE HORIZONTAL LINE PASSING THROUGH THE POINT AT WHICH THE VENTRICULAR DEPOLARIZATION IS ENDED (THE J POINT) AND THE LEVEL OF THE T-P SEGMENTS IS ABOVE THE ZERO p.l. IN LEAD I, AND BELOW THE ZERO p.l. IN LEAD III, ONE MAY CONCLUDE THAT THERE IS A "CURRENT OF INJURY" IN THE REPOLARIZED HEART WHICH HAS A VECTOR DIRECTED FROM:
    A. Right to left
    B. Left to right
    C. Head to foot   Ref. p. 191

56. THE VECTOR OF THE CURRENT OF INJURY POINTS ITS POSITIVE END TOWARD THE:
    A. Normal fully repolarized portion of the heart
    B. Ischemic partly depolarized portion of the heart
    C. Necrotic (infarcted) portion of the heart
    D. S-A node
    E. A-V node   Ref. p. 191

57. IF A STRONG CURRENT OF INJURY IS PREDOMINANT IN V-3 (J POINT 12 MILLIVOLTS ABOVE THE T-P SEGMENT) IT COULD BE STRONGLY SUGGESTED THAT THERE IS AN INJURED ZONE OF MYOCARDIUM IN THE:
    A. Base of the heart (diaphragmatic side of the right ventricle)
    B. Posterior wall of the left ventricle
    C. Left lateral wall of the left ventricle
    D. Posterior portion of the interventricular septum
    E. Anterior wall of the right or left ventricle near the septum
                                           Ref. p. 192

58. A VERY PROLONGED DEPOLARIZATION OF THE VENTRICULAR APEX WILL TEND TO BE FOLLOWED BY REPOLARIZATION WHOSE VECTOR FORCES:
    A. Are very rapid
    B. Have the same direction as the mean depolarization vectors (positive same side)
    C. Are very slow
    D. Have the opposite direction as the mean depolarization vectors
    E. Will precede the atrial repolarization   Ref. p. 194

SECTION I - THE CARDIOVASCULAR SYSTEM

QUESTIONS 59-69 (INCLUSIVE) SHOULD BE ANSWERED ACCORDING TO THE FOLLOWING METHOD:
If only sentence 1 is correct mark your choice with the letter A
If sentences 1 and 2 are the only correct ones, mark your choice with the letter B
If sentences 1, 2 and 3 are correct, mark your choice with the letter C
If sentences 1, 2, 3 and 4 are correct, mark your choice with the letter D
If sentences 1, 2, 3, 4 and 5 are correct, mark your choice with the letter E

59. ATRIOVENTRICULAR BLOCK MAY BE CAUSED BY:
   1. Ischemia of the atrioventricular nodal fibers
   2. Compression of the A-V node by scarring
   3. Inflammatory changes of the A-V node (rheumatic carditis for example)
   4. Vagal stimulation
   5. An infusion of acetylcholine   Ref. p. 196

60. THE SO-CALLED STOKES-ADAMS SYNDROME IS CHARACTERIZED BY:
   1. Transient, episodic complete A-V block
   2. Ventricular pacemaking in the A-V node
   3. Temporary cerebral hypoxia
   4. Compensatory increased cardiac output
   5. Hypersensitive carotid baroreceptors   Ref. p. 197

61. VENTRICULAR PREMATURE BEATS:
   1. Give rise to prolonged QRS complexes
   2. Are followed by repolarization vectors which are opposite in direction to depolarization vectors
   3. Are invariably correlated with non-pathologic factors, such as an excess of coffee, smoking, and do not pose any danger
   4. Which are positive in lead I and negative in lead III have a vector whose positive end is directed down and toward the right ventricle
   5. Will not appear around infarcted areas because myocardial necrosis prevents an area around it from acting as a source of depolarization waves
   Ref. pp. 198-199

62. DIAGNOSTIC CRITERIA FOR THE DIAGNOSIS OF A NODAL TACHYCARDIA ARE:
   1. QRS complexes of normal configuration
   2. QRS complexes of normal duration
   3. QRS complexes not preceded by P waves
   4. Relatively slow rates of ventricular contraction (below 70 beats/minute)
   5. A brief P wave followed shortly after by a QRS with a "nodal" configuration   Ref. pp. 199-200

63. THE PHYSIOLOGICAL PARAMETERS FAVORABLY MODIFIED BY QUINIDINE IN THE PREVENTION OF VENTRICULAR FIBRILLATION (WHICH MAY EVOLVE FROM A BOUT OF VENTRICULAR TACHYCARDIA) ARE THE:
   1. Prolongation of the refractory period of ventricular muscle
   2. Speeding up of the conduction of the depolarization wave
   3. Uniformity introduced in the rate of discharges from the A-V node
   4. Acceleration of normal atrial ventricular conduction
   5. Lowering of the potassium conductance
   Ref. p. 200

SECTION I - THE CARDIOVASCULAR SYSTEM

64. ATRIAL FIBRILLATION IS PHYSIOLOGICALLY CHARACTERIZED BY:
    1. A very irregular constantly changing atrial pathway for the depolarization wave
    2. Low atrial voltages in the electrocardiogram
    3. QRS of normal configuration
    4. QRS complexes occurring at irregular intervals
    5. Lower cardiac output than with normal sinoatrial rhythm
       Ref. pp. 200-201

65. BLOOD VISCOSITY:
    1. Increases exponentially with increasing hematocrits
    2. Decreases paradoxically in tubes of very small diameter and does not have a constant value in glass viscometers of different calibers
    3. Decreases with higher rates of volume flow of blood when the initial flow is near zero
    4. Does not change significantly with temperature variations between 20° C and 37° C
    5. Is greatly affected by small variations in plasma protein concentrations
       Ref. p. 205

66. RESISTANCE TO BLOOD IS KNOWN TO:
    1. Increase exponentially in progressively narrower vessels
    2. Increase in cases of turbulent flow
    3. Decrease when high Reynold's numbers are obtained
    4. Be low in each capillary of a large capillary bed
    5. Be low in polycythemic blood   Ref. p. 209

67. THE TENDENCY TO REACH A HIGH REYNOLDS NUMBER IS HIGHER WHEN THE:
    1. Mean linear velocity of blood flow is high
    2. Radius of the vessel is small
    3. Density of the blood is low
    4. Viscosity of blood is high   Ref. p. 209

68. IT WOULD BE CORRECT TO SAY THAT:
    1. A mean pressure in the arteries equal to 100 mm Hg is the same as that of 136 cms of water
    2. A mean pressure equal to 100 mm Hg equals that of 128 cms of blood
    3. Veins of a person whose standing height from heart to toes equals 128 cms have 100 mm Hg in their toes' venous blood, plus the pressure remaining in the venules after blood has passed the capillaries.
    4. Venous blood in the toes of such a standing person would be unable to return to the heart
    5. Capillaries in the toes of such a standing person have a pressure less than 100 mm Hg   Ref. pp. 210, 225

69. THE CRITICAL CLOSING PRESSURE:
    1. Is higher when sympathetic fibers to arteriolar smooth muscle are stimulated maximally
    2. Usually equals less than 2 mm Hg intraarteriolar pressure
    3. Is a value of pressure above which there is no blood perfusion
    4. Has a fixed value regardless of the sympathetic tone
    5. Should have a low value during vasoconstriction
       Ref. pp. 213, 326

SECTION I - THE CARDIOVASCULAR SYSTEM

SELECT THE ONE MOST APPROPRIATE ANSWER:

70. INHIBITION OF SYMPATHETIC NEUROTRANSMITTERS REACHING THE ARTERIOLAR SMOOTH MUSCLE SHOULD RESULT IN:
    A. Decrease of blood flow in the capillary bed distal to the arterioles
    B. Increase of the blood flow in the capillaries distal to the arterioles
    C. Increase of the critical closing pressure
    D. Decrease the arteriolar radius
    E. Increase central aortic pressure   Ref. p. 213

71. THE COMPLIANCE OF A CERTAIN LENGTH OF THE VENA CAVA:
    A. Is greater than a similar length of the aorta
    B. Is the reciprocal of its distensibility
    C. Is about the same as that of the pulmonary artery
    D. Has an invariable numerical value regardless of the sympathetic tone
                                                 Ref. p. 214

72. FROM COMPLIANCE CURVES OBTAINED FOR LARGE ARTERIES AND LARGE VEINS ONE CAN CONCLUDE THAT FOR A GIVEN INCREMENT IN INTRALUMINAL PRESSURE THE CHANGES IN THE VESSELS CONTENT WILL BE:
    A. Relatively larger for the aorta
    B. Relatively larger for the vena cava
    C. Essentially the same in both vessels   Ref. Fig. 18-15, p. 214

73. ALL OF THE FOLLOWING STATEMENTS CONCERNING CAPILLARY CIRCULATION ARE TRUE, EXCEPT TO SAY THAT:
    A. Capillaries together account for the largest surface area of all vascular compartments (arteries, veins, capillaries)
    B. Red blood cells move within capillaries at the slowest speed found in circulatory beds
    C. Capillaries have a very thin wall because this wall sustains the minimal tangential tension encountered in all vessels
    D. Capillaries do not respond by active constriction to local epinephrine
    E. Capillaries together contain more blood than that contained in the veins
                                                 Ref. p. 219

74. FRICTION ACCOUNTS FOR MOST OF THE ENERGY LOSS OF CIRCULATING BLOOD. THIS LOSS IS MINIMAL IN THE:
    A. Capillaries of the systemic circulation
    B. Arterioles
    C. Pulmonary arterioles
    D. Aorta   Ref. p. 219

75. THE GREATEST FLOW OF BLOOD OUT OF THE ARTERIES AND INTO THE CAPILLARY SYSTEM (PERIPHERAL RUN-OFF) TAKES PLACE DURING:
    A. Iso-volumetric contraction phase
    B. Early ejection phase
    C. Last part of diastole
    D. Atrial systole   Ref. p. 220

SECTION I - THE CARDIOVASCULAR SYSTEM

76. A GREATER PULSE PRESSURE IS POSITIVELY CORRELATED WITH A:
    A. Large systolic stroke volume
    B. Higher compliance of the arterial walls
    C. Higher frequency of contraction of the heart, with a constant cardiac output
    D. Prolonged period of systolic ejection   Ref. p. 220

77. A HYPERTENSIVE PATIENT WILL USUALLY HAVE A LARGE PULSE PRESSURE:
    A. Even with a normal systolic stroke volume
    B. Only with a very large systolic stroke volume
    C. Due to a high compliance of the arterial walls in hypertension
    D. Because the heart is usually hypertrophic
                                                    Ref. p. 220

78. AGING CAUSES CHANGES IN THE ARTERIAL COLLAGEN WHICH RESULT IN:
    A. Decreased compliance of the arterial walls
    B. Increased compliance of the arterial walls but slower pulse velocity
    C. Low pulse pressure
    D. A diminished mean arterial pressure   Ref. pp. 220-221

79. THE MOST RELIABLE INDEX OF THE VASCULAR RESISTANCE IMPOSED UPON THE HEART IN THE ELDERLY PATIENT IS THE:
    A. Level of diastolic pressure
    B. Magnitude of the pulse pressure
    C. Level of the systolic pressure
    D. Rate of systolic pressure rise
    E. Velocity of pulse wave formation   Ref. p. 221

80. HEMODYNAMIC CHANGES FOUND IN CASES OF PATENT DUCTUS ARTERIOSUS (PDA) INCLUDE ALL OF THE FOLLOWING FEATURES WITH ONE EXCEPTION, AND IT WOULD BE WRONG TO INCLUDE THE STATEMENT THAT PDA:
    A. Has a rapid runoff into the pulmonary artery
    B. Its high systolic pressure is largely due to an increased systolic stroke volume
    C. Has a large pulse pressure
    D. Has a relatively elevated diastolic pressure
    E. Has an aortic pressure tracing with the incisura placed nearer the diastolic level than in normal cases   Ref. p. 221

81. THE VELOCITY OF TRANSMISSION OF THE PULSE WAVE IN THE AORTA IS:
    A. Greater when the compliance decreases with age
    B. Greater in the young people than among older persons
    C. Equals the velocity of flow of red cells in the center of the parabolic front of cells flowing in the aorta
    D. Is reciprocally related to the cardiac output
                                                    Ref. p. 221

SECTION I - THE CARDIOVASCULAR SYSTEM

82. IF THE PRESSURE IS MEASURED, WITH A HIGHLY SENSITIVE HIGH FIDELITY SYSTEM AT THE LEVEL OF THE ROOT OF THE AORTA AND AT SEVERAL LEVELS OF THE ARTERIAL SYSTEM PROGRESSIVELY REMOVED FROM THE HEART, ONE FINDS THAT THE "PRESSURE PULSE CONTOUR" CHANGES. WHICH OF THE FOLLOWING STATEMENTS IS THE ONLY ONE IN ERROR?:
    A. The systolic pressure rises more and more in distal areas
    B. The diastolic pressure tends to be slightly lower in distal areas
    C. The pulse pressure is increased in distal areas
    D. The time spent by the increment of pressure diminishes (shorter systole) in distal areas
    E. The mean pressure is gradually increased in distal areas
    Ref. Fig. 19-6, p. 222

83. OF ALL THE POSSIBLE REASONS LISTED BELOW, WHICH ONE SEEMS TO EXPLAIN BEST THE PHENOMENA LISTED IN THE PREVIOUS QUESTIONS?:
    A. Diminished compliance of the arteries in distal portions of the arterial tree
    B. An artifact due to the instruments used which can never match the periodicity of the arterial system
    C. Increased momentum of blood as it moves down the arterial tree
    D. Gravity forces   Ref. p. 222

84. WHEN ARTERIOLES ARE SEVERELY CONSTRICTED, THE CAPILLARIES DISTAL TO THE ARTERIOLE ARE PERFUSED BY A:
    A. Very rapid flow
    B. Pulsatile flow
    C. Non-pulsatile sluggish flow
    D. Non-viscous newtonian flow   Ref. p. 223

85. FROM A KNOWLEDGE OF LAPLACE'S LAW IT WOULD BE CORRECT TO PREDICT THAT, BECAUSE THE PRESSURE IS HIGHER IN CAPILLARIES THAN IN VEINS, THE TANGENTIAL TENSION EXERTED ON THE CAPILLARIES' WALLS SHOULD BE IN COMPARISON TO THAT EXERTED ON VEINS:
    A. Greater (in capillaries)
    B. Lesser (in capillaries)
    C. Equal (in capillaries)   Ref. p. 224

86. CENTRAL VENOUS PRESSURE IS THE VALUE OF BLOOD PRESSURE FOUND IN THE:
    A. Center of the superior vena cava
    B. Center of the inferior vena cava
    C. Mean right ventricular pressure
    D. Right atrial pressure
    E. Pressure in the central vein   Ref. p. 224

87. IF THE RIGHT VENTRICLE UNDERGOES THE METABOLIC CHANGES THAT LEAD TO WHAT IS KNOWN AS RIGHT (VENTRICULAR CONGESTIVE) FAILURE, THE CENTRAL VENOUS PRESSURE WILL:
    A. Rise
    B. Diminish
    C. Decrease during a fist stage, then fluctate minimally with respiration
    D. Remain the same   Ref. p. 224

SECTION I - THE CARDIOVASCULAR SYSTEM

88. THE VARIOUS FACTORS ENUMERATED BELOW TEND TO <u>INCREASE</u> THE CENTRAL VENOUS PRESSURE (CVP) WITH ONE EXCEPTION; IT WOULD BE <u>WRONG</u> TO INCLUDE AMONG THESE FACTORS A STATEMENT SAYING THAT CVP IS INCREASED BY A(N):
   A. Increased blood volume
   B. Increase in sympathetic tone to the venous system
   C. Dilatation of arterioles in the systemic circuit
   D. Improved efficiency of the contraction of a previously failing right ventricle   Ref. p. 224

89. THE SUBCLAVIAN VEIN OF A MAN WHOSE ARM (60 CMS LONG) IS HANGING DOWN PASSIVELY NEXT TO HIS THORAX IS FREQUENTLY COLLAPSED. BLOOD RETURNING TO THE HEART FROM THE FINGERS OF THE HAND WOULD HAVE A VENOUS PRESSURE IN THE HAND EQUAL TO APPROXIMATELY:
   A. 0 mm Hg
   B. 5 mm Hg   HINT: $60 \times 1.060 = \chi \times 13.59$
   C. 45 mm Hg
   D. 250 mm Hg   Ref. p. 225

90. IF SUCH A MAN RAISES HIS ARM HORIZONTALLY, VENOUS PRESSURE IN THE VEINS OF THE HAND WILL BE APPROXIMATELY:
   A. 0 mm Hg
   B. 5 mm Hg
   C. 25 mm Hg
   D. 250 mm Hg   Ref. p. 225

91. VEINS WITHIN THE THORAX AND THE CRANIAL CAVITY ARE, AS OPPOSED TO EXTERNAL JUGULAR OR THE SUBCLAVIAN:
   A. Open all the time
   B. Collapsed all the time
   C. Sometimes open, sometimes collapsed, depending on the position of the body
   D. Compressed by viscerae in the immediate vicinity
                       Ref. p. 225

92. THE DISTANCE BETWEEN THE RIGHT ATRIUM AND ANGLE OF THE MANDIBLE IS ABOUT 25.0 CMS IN AN ADULT HUMAN. NORMALLY, THE EXTERNAL JUGULAR VEIN, AT THE LEVEL OF THE ANGLE OF THE MANDIBLE, IN A PERSON SITTING UP IN BED, IS COLLAPSED. THEREFORE ITS INTRAVENOUS PRESSURE AT THAT LEVEL IS ZERO. HOWEVER, IN PATIENTS WITH INSUFFICIENCY OF RIGHT VENTRICULAR CONTRACTILITY (CONGESTIVE FAILURE OF THE RIGHT VENTRICLE) THE EXTERNAL JUGULAR IS OFTEN DISTENDED UP TO SUCH A HIGH LEVEL IN THE SITTING POSITION. THE VENOUS PRESSURE AT THE RIGHT ATRIUM IS THEN APPROXIMATELY:
   A. 2.0 mm Hg
   B. 20.0 mm Hg
   C. 50.0 mm Hg
   D. 100.0 mm Hg

Hint: $\text{Height}_{blood} \times \text{density}_{blood} = \text{height}_{mercury} \times \text{density}_{mercury}$
                       Ref. p. 225

SECTION I - THE CARDIOVASCULAR SYSTEM

93. PERMEABILITY OF CAPILLARIES IN THE LEGS AND FEET OF HUMANS COULD ALLOW, IF SUBJECTED TO PRESSURES EQUAL TO 100 mm Hg (PERSON STANDING WITHOUT MUSCULAR MOVEMENTS), A LOSS OF CIRCULATING BLOOD VOLUME EQUAL TO:
    A. 0%
    B. 1.0%
    C. 20%
    D. 60%   Ref. p. 226

94. ALL PRESSURES WITHIN THE CIRCULATORY SYSTEM SHOULD BE COMPARED TO A "REFERENCE POINT" WHERE THERE IS THE LEAST VARIABILITY IN GRAVITATIONAL FORCES EVEN UNDER THE MOST DIVERSE CIRCUMSTANCES. THIS POINT IS LOCATED IN THE IMMEDIATE VICINITY OF THE:
    A. Superior vena cava
    B. Inferior vena cava immediately below the diaphragm
    C. Pulmonary capillaries
    D. Left atrium at the level of the mitral valve
    E. Right atrium at the level of the tricuspid valve
                                                        Ref. p. 227

95. BECAUSE OF THE CALIBER OF CAPILLARIES IN HUMANS ONE CAN EXPECT THAT RED BLOOD CELLS MAY PASS THROUGH THEM:
    A. In groups of 2-5, side by side (40 microns in diameter)
    B. In groups of 10, side by side (80 microns in diameter)
    C. One by one, in a single file (8 microns or less)
    D. As a rather large compact column (200 microns in diameter)
                                                        Ref. p. 230

96. CAPILLARY PORES ARE IN MUSCULAR CAPILLARY BEDS (TONGUE, SKELETAL MUSCLE):
    A. Very abundant, and constitute about 50% of the surface area
    B. Very few, far apart and constitute less than 1% of the surface area
    C. Can open up according to need, very rapidly, providing rapid increases in areas with high diffusibility
    D. Non-existent   Ref. p. 230

97. PRECAPILLARY SPHINCTERS OPEN AND CLOSE REGULATING ACTUAL CAPILLARY PERFUSION AND DIVERTING PREDOMINANT FLOW THROUGH META-ARTERIOLES OR THROUGH CAPILLARIES. THE FACTOR WHICH BEST CORRELATES WITH OPENING PRECAPILLARY SPHINCTERS IS THE:
    A. Relative hypoxia of tissues
    B. Relative hypertension of the arterioles
    C. Venous pressure
    D. Concentration of albumin in plasma   Ref. p. 231

98. ALL THE FOLLOWING FACTORS FAVOR THE DIFFUSION OF A SUBSTANCE FROM THE LUMEN OF A CAPILLARY TO THE CELLS IN THE IMMEDIATE VICINITY, WITH THE EXCEPTION OF ONE. IT WOULD NOT BE CORRECT TO SAY THAT DIFFUSION IS AIDED BY:
    A. The concentration gradient of the substance
    B. The thermal motion of the particles
    C. The area available for diffusion
    D. The hydrostatic pressure of the fluid in which the particles are dissolved
    E. A different solubility (between diffusible substance and membrane components)   Ref. p. 232

SECTION I - THE CARDIOVASCULAR SYSTEM

99. THE RATE OF VOLUME FLOW OF PLASMA WATER THROUGH THE LENGTH OF A GIVEN CAPILLARY (CAPILLARY FLOW) MAY BE COMPARED WITH THE RATE OF WATER DIFFUSION THROUGH THE FEW PORES AVAILABLE IN THE CAPILLARY WALL (WATER DIFFUSION). IT IS PROBABLY TRUE THAT CAPILLARY FLOW:
    A. Equals water diffusion
    B. Is slightly greater than water diffusion
    C. Is many times smaller than water diffusion
    Ref. p. 232

100. DIFFUSION OF CARBON DIOXIDE AND OXYGEN THROUGH CAPILLARY WALLS IS KNOWN TO BE GREATER THAN DIFFUSION OF SODIUM CHLORIDE, UREA AND GLUCOSE. THIS IS MOST LIKELY DUE TO THE FACT THAT THE TWO GASES:
    A. Have a smaller molecular size than the water soluble solutes
    B. Are more water soluble than the solutes mentioned above
    C. Are lipid soluble
    D. Have electrical properties that make them more diffusible
    E. Have a greater concentration gradient than the water soluble solutes
    Ref. p. 232

101. PINOCYTOSIS IS A PROCESS:
    A. By which phagocytes in the endothelial wall ingest colloid particles present in the lumen
    B. Whereby the bulk of transport of crystalloids occur
    C. Which probably plays a small part in transendothelial diffusion and transport
    D. Taking place mainly next to capillary pores
    Ref. p. 233

102. BECAUSE OF THE DONNAN EQUILIBRIUM, THE CONCENTRATION OF CATIONS IN THE SAME SIDE WHERE PROTEIN CONCENTRATION IS GREATER (COMPARING PLASMA VERSUS EXTRAVASCULAR SPACE) IS:
    A. Equal to that of crystalloid anions on the same side of the membrane
    B. Much greater than that of crystalloid anions on the same side
    C. Much less than that of crystalloid anions on the same side
    D. Maintained only through an active transport mechanism that opposes rapid outflux of cations
    Ref. p. 236

103. DUE TO THE INEQUALITY OF THE DISTRIBUTION OF CATIONS ACROSS CAPILLARY MEMBRANES, THE CONTRIBUTION OF CRYSTALLOID CATIONS WITHIN THE INTRAVASCULAR COMPARTMENT TO THE COLLOID OSMOTIC PRESSURE (ONCOTIC PRESSURE) IS:
    A. Negligible
    B. Overwhelming
    C. About 50% of the total oncotic pressure
    D. About 1/3 of the total oncotic pressure
    Ref. p. 236

SECTION I - THE CARDIOVASCULAR SYSTEM

104. IF ONE WERE TO ACCEPT VALUES OBTAINED BY THE ISOVOLU-METRIC OR THE ISOGRAVIMETRIC METHODS OF MEASURING INTRA-CAPILLARY PRESSURE IN LIVING MAMMALIAN TISSUES (PLACED AT THE SAME LEVEL AS THE HEART) RATHER THAN THOSE VALUES OB-TAINED BY DIRECT INTRODUCTION OF MICROPIPETTES CONNECTED TO MANOMETERS, CAPILLARY PRESSURE VALUES ARE APPROXI-MATELY EQUAL TO:
    A. 60 mms Hg
    B. 25 cms water
    C. 25 mms Hg
    D. 17 mms Hg
    E. 5 mms Hg   Ref. p. 234

105. RAPID WATER FLUXES BETWEEN CAPILLARY LUMINA AND THE INTERSTITIAL SPACE ARE DEPENDENT ON OSMOTIC FORCES PRIMARILY DETERMINED BY THE CONCENTRATION OF:
    A. Sodium and chloride ions
    B. Gamma-globulin molecules
    C. Fibrinogen molecules
    D. Albumin molecules   Ref. p.

106. THE VALUE OF TOTAL OSMOTIC PRESSURE THAT CONTRIBUTES TO THE CONTROL OF WATER FLUXES BETWEEN INTRACELLULAR CYTO-PLASM AND EXTRACELLULAR (INTERSTITIAL) SPACES IS, WHEN COM-PARED TO THE COLLOID OSMOTIC PRESSURE OF PLASMA:
    A. Equal in value to the latter
    B. Smaller than the latter
    C. Slightly greater than the latter
    D. Very much greater than the latter   Ref. p. 236

107. EDEMA CAN OCCUR AS A RESULT OF ALL OF THE FOLLOWING FAC-TORS, WITH ONE EXCEPTION. THE FACTOR INCAPABLE OF PROMOT-ING EDEMA IS A(N):
    A. Increase in mean capillary pressure from 17 mm Hg to 37 mm Hg
    B. Increase in the negative pressure normally found in the interstitial fluid space
    C. Decrease in albumin concentration in circulating plasma
    D. Obstruction of lymphatic vessels
    E. Increased vasomotion with persistent tonic contraction of precapillary sphincters   Ref. p. 238

108. THE REASON WHY MACROMOLECULES RETURNING TO THE GENERAL CIRCULATION THROUGH LYMPHATIC VESSELS CANNOT "ESCAPE" FROM THE LYMPHATIC LUMEN ONCE THEY HAVE ENTERED INTO THESE VESSELS IS THAT:
    A. Lymphatics have no pores in their entire length
    B. Pressure is always negatively exerted on lymphatic contents
    C. Macromolecules are always larger than lymphatic pores
    D. Terminal portions of lymphatics have openings arranged so that a "flap valve" action permits a unidirectional flow of macromolecules
                                             Ref. p. 242

SECTION I - THE CARDIOVASCULAR SYSTEM

109. THE RATE OF LYMPH FLOW WILL BE ACCELERATED BY ALL THE FACTORS QUOTED BELOW, WITH ONE EXCEPTION. THUS, IT IS MOST LIKELY THAT THERE WILL BE <u>SLOWER</u> LYMPHATIC FLOW WHEN THERE IS:
    A. A hydrostatic intracapillary pressure equal to 35 mm Hg in the venular end
    B. An interstitial fluid protein concentration equal to about 9.5 gms%
    C. Increased permeability of the capillary walls
    D. Scarring of the lymph nodes   Ref. p. 243

110. EDEMA OF THE INTERSTITIAL SPACES WILL BE FAVORED BY A(N):
    A. Increase of interstitial pressure from subatmospheric to slightly above atmospheric pressure
    B. Increase in the concentration of albumin in circulating plasma
    C. Lowering of central venous pressure from 25 to 5 mms Hg
    D. Decrease in pressure from above atmospheric to subatmospheric pressure within the interstitial space
    Ref. p. 246

111. WITH ONE EXCEPTION, CAPILLARY PRESSURE INCREASES IN THOSE CONDITIONS LISTED BELOW. NAME THE <u>EXCEPTION</u>:
    A. Congestive heart failure
    B. Thrombosis of veins
    C. Arteriolar dilatation due to local release of histamine
    D. Arteriolar dilatation in reactive hyperemia
    E. Assumption of supine (decubitus) position
    Ref. p. 248

112. COLLOID OSMOTIC PRESSURE CAN BE LOWERED WITHOUT CAUSING EDEMA PROVIDED THE DROP IN PRESSURE DOES NOT EXCEED:
    A. 1 mm Hg
    B. 7 mm Hg
    C. 25 mm Hg
    D. 45 mm Hg   Ref. p. 248

113. THE CUBOIDAL CELLS OF THE CHOROID PLEXUS ARE RESPONSIBLE FOR THE DIFFERENCE BETWEEN CEREBROSPINAL FLUID AND INTERSTITIAL FLUID FORMED BY ULTRAFILTRATION OF PLASMA FROM CAPILLARIES. THIS IS BECAUSE CELLS OF THE CHOROID PLEXUS "SECRETE" A RELATIVE EXCESS OF:
    A. Protein anions
    B. Potassium ions
    C. Chloride ions
    D. Sodium ions
    E. Glucose molecules   Ref. p. 253

114. WITH RESPECT TO PLASMA OSMOLARITY, THE OSMOTIC CONCENTRATION OF NEWLY FORMED CEREBROSPINAL FLUID IN CEREBRAL VENTRICLES (LATERAL, 3rd AND 4th) IS:
    A. Higher
    B. Lower
    C. Equal   Ref. p. 253

SECTION I - THE CARDIOVASCULAR SYSTEM

115. BECAUSE OF THE SPECIFIC DIFFUSION RATES BETWEEN BRAIN SUBSTANCE AND SUBARACHNOID CEREBROSPINAL FLUID, AN ANALYSIS OF THE LATTER WOULD GIVE A REASONABLY FAIR REPRESENTATION OF THE CONCENTRATION WITHIN THE BRAIN OF:
    A. Sucrose
    B. Protein
    C. Alcohol
    D. Penicillin
    E. Streptomycin   Ref. p. 253

116. PROTEIN MOLECULES WHICH HAVE LEAKED OUT OF BRAIN CAPILLARIES RETURN TO THE CIRCULATION BY PASSAGE THROUGH:
    A. Veins
    B. Perivascular spaces
    C. Cerebral lymphatics
    D. Capillaries of the dura   Ref. p. 254

117. A CASE OF SO-CALLED NON-COMMUNICATING HYDROCEPHALUS MAY BE CAUSED BY A BLOCK OF THE:
    A. Choroid plexus
    B. Arachnoidal villi due to chronic leptomeningitis
    C. Aqueduct of Sylvius
    D. Entrance to the spinal canal   Ref. p. 255

118. IF THE PRESSURE IN THE SPINAL CANAL IS FOUND TO BE EQUAL TO 300 mm OF WATER AND THE PRESSURE IN THE CISTERNA MAGNA IS FOUND TO BE 130 mm WATER IN A PERSON LYING ON HIS SIDE, ONE CAN CONCLUDE THAT:
    A. There is greater rate of formation of spinal fluid in the spinal canal's perivascular arachnoidal spaces than at the ventricular's choroid plexuses
    B. There is blockage of the normal circulation of the cerebrospinal fluid from the spinal canal upwards to the cerebral surface
    C. This is a normal value for both pressures
    D. Reabsorption of fluid is much faster in the cerebral than in the spinal arachnoid surface in the person lying on his side
                                                      Ref. p. 255

119. IF THERE IS COMPLETE BLOCKAGE OF THE CEREBROSPINAL FLUID CIRCULATION IN A HUMAN, A MANEUVER WHICH INCREASES THE VENOUS PRESSURE OF THE CEREBRAL VEINS (QUECKENSTEDT'S TEST) WILL CAUSE:
    A. A rapid exophthalmos
    B. A rapid increase in papilledema
    C. A rapid drop in arterial blood pressure
    D. No increase of spinal fluid pressure
    E. Rapid increase of spinal fluid pressure
                                                      Ref. p. 255

SECTION I - THE CARDIOVASCULAR SYSTEM

120. THE AREA OF THE BRAIN WHERE THE BLOOD-BRAIN BARRIER SEEMS TO BE MOST PERMEABLE IS PROBABLY LOCATED IN THE:
    A. Frontal and parietal cortex
    B. Hypophysis
    C. Cerebellum
    D. Brain stem
    E. Hypothalamus   Ref. p. 256

121. THE AVERAGE NORMAL INTRAOCULAR PRESSURE AS MEASURED IN THE ANTERIOR CHAMBER OF THE EYE EQUALS ABOUT:
    A. 1 mm Hg
    B. 5 mms Hg
    C. 20 mms Hg
    D. 100 mms Hg
    E. 10 cms water   Ref. p. 257

122. THE FORMATION OF AQUEOUS HUMOR OF THE EYE IS BELIEVED TO BE DUE TO:
    A. Diffusion of water from the capillaries of the ciliary body due to a hydrostatic pressure gradient between capillaries and the anterior chamber of the eye
    B. Active secretion of sodium ions by cells of the ciliary processes followed by osmotic water transfer from capillaries into the anterior chamber of the eye
    C. Dilution of the vitreous humor with plasma water
    D. Dialysis of the vitreous humor   Ref. p. 256

123. IN THE MAJORITY OF PATIENTS AFFLICTED BY GLAUCOMA, THE INCREASED PRESSURE WITHIN THE EYE'S CHAMBERS IS DUE TO:
    A. Increased formation of intraocular fluid
    B. Decreased reabsorption through the canal of Schlemm
    C. Increased pressure in the capillaries of the eye
    D. Thrombosis of venules in retinal plexuses
                                    Ref. p. 258

124. DUE TO A SERIES OF ANATOMICAL AND PHYSICAL FACTORS, THE TENDENCY FOR EXTRACELLULAR FLUID TO ACCUMULATE IN A "POTENTIAL" SPACE IN THE BODY (PLEURA, PERITONEUM) IS HIGHER IN THE:
    A. Pleura
    B. Pericardium
    C. Peritoneum
    D. Synovial cavities
    E. Bursae   Ref. p. 259

125. ASSUMING AN EQUAL SYSTOLIC STROKE VOLUME FOR THE RIGHT VENTRICLE AS FOR THE LEFT VENTRICLE, AND KNOWING THE RELATIONS BETWEEN COMPLIANCE AND PRESSURES CAUSED BY SYSTOLIC FILLING OF THE PULMONARY ARTERIES AND THE AORTA, ONE MAY CONCLUDE THAT THE RATIO $\frac{\text{PULMONARY COMPLIANCE}}{\text{AORTIC COMPLIANCE}}$ EQUALS:
    A. 1.0
    B. 5.0
    C. 1/5
    D. 25
    E. 1/25   Ref. p. 262

SECTION I - THE CARDIOVASCULAR SYSTEM

126. AN INCREASE OF LEFT VENTRICULAR END <u>DIASTOLIC</u> PRESSURE WILL BE ASSOCIATED WITH A RISE IN LEFT ATRIAL AND PULMONARY CAPILLARY PRESSURE, AS LEFT VENTRICLE, LEFT ATRIUM AND PULMONARY VEINS COMMUNICATE FREELY DURING VENTRICULAR DIASTOLE. SUCH AN INCREASE IN PRESSURE TO A LEVEL EQUAL TO 15-30 mm Hg IS INDICATIVE OF:
    A. Pulmonary artery hypertension
    B. Right ventricular failure
    C. Left ventricular failure
    D. Increased compliance of the pulmonary artery
    Ref. p. 263

127. IF CARDIAC OUTPUT IS INCREASED THE PRESSURE WITHIN THE PULMONARY CIRCUIT MAY RISE. DUE TO THE DISTENSIBILITY OF THE PULMONARY VESSELS, PULMONARY HYPERTENSION WILL TAKE PLACE TO A SIGNIFICANT DEGREE ONLY WHEN SUCH AN INCREMENT IN CARDIAC OUTPUT EXCEEDS THE NORMAL BY:
    A. One-fold
    B. Two-fold
    C. Four-fold
    D. Six-fold   Ref. p. 264

128. IF A LOCAL REGIONAL AREA OF THE PULMONARY VESSELS IS PERFUSED BY BLOOD WHICH HAS DISTINCTLY LOWERED OXYGEN CONTENT, THE VESSELS OF SUCH AREA WILL:
    A. Dilate
    B. Constrict
    C. Undergo a reactive hyperemia
    D. Thrombose   Ref. p. 265

129. EXTENSIVE OCCLUSION OF PULMONARY VESSELS BY THROMBI ORIGINATED IN THE PELVIS OR IN THE LOWER EXTREMITIES MAY RESULT IN:
    A. Left ventricular failure
    B. Right ventricular failure
    C. Increased cardiac output
    D. Complete compensation by excessive distensibility of non-occluded vessels   Ref. p. 265

IN THE FOLLOWING QUESTIONS (130-159) A STATEMENT IS FOLLOWED BY FOUR POSSIBLE ANSWERS. ANSWER BY USING THE KEY BELOW:
A. If only 1, 2 and 3 are correct
B. If only 1 and 3 are correct
C. If only 2 and 4 are correct
D. If only 4 is correct
E. If all are correct

130. EMPHYSEMA INCREASES RESISTANCE TO BLOOD FLOW THROUGH PULMONARY CAPILLARIES BY:
    1. Diminution of total number of vessels
    2. Vasoconstriction due to relative generalized hypoxia
    3. Vasoconstriction due to alveolar hypoventilation and local hypoxia
    4. Increase in cardiac output due to generalized hypoxia
    Ref. p. 266

SECTION I - THE CARDIOVASCULAR SYSTEM

131. IN DIFFUSE SCLEROSIS OF THE LUNGS:
   1. The pulmonary arterial pressure may be normal at rest
   2. The right ventricle never undergoes failure if given the chance to undergo hypertrophy
   3. Pulmonary arterial pressure rises rapidly even with mild exercise
   4. Gas diffusion is unimpeded   Ref. p. 266

132. IN ATELECTASIS:
   1. A bronchi is plugged
   2. The corresponding alveolus is collapsed
   3. Alveolar air is reabsorbed by the blood still perfusing the collapsed alveoli
   4. Blood flow is increased through atelectatic alveoli, because the diminished elastic recoil now ceases to constrict the pulmonary vessels
   Ref. p. 266

133. THE SO-CALLED PULMONARY "WEDGE PRESSURE":
   1. Measures potential minus kinetic energy in pulmonary capillaries
   2. Measures more closely events in the pulmonary artery
   3. Measures more closely the pressure within the left atrium
   4. Can be used to estimate the end systolic right ventricular pressure
   Ref. p. 267

134. THE REASON WHY PULMONARY EDEMA DOES NOT DEVELOP EASILY IN CHRONIC ELEVATIONS OF PULMONARY CAPILLARY PRESSURES IS DUE TO THE:
   1. High oncotic pressure in the pulmonary blood
   2. Lower diffusibility of water through pulmonary capillary walls
   3. Negative interstitial pressure in the lungs
   4. Extremely rapid run-off of fluid from pulmonary interstitial spaces into the enlarged pulmonary lymphatics
   Ref. p. 268

135. THE DISTRIBUTION OF BLOOD FLOW IN VARIOUS ORGANS IS SET IN SUCH A FASHION THAT BLOOD PERFUSING THE:
   1. Brain is very largely in excess of its metabolic needs
   2. Liver is proportional to its metabolic needs
   3. Skin is only determined by the surface area and not by physical changes of the environment
   4. Kidney is very largely in excess of its metabolic needs
   Ref. p. 270

136. BLOOD FLOW THROUGH:
   1. Skeletal muscles increases proportionately with the increase in the rate of metabolism
   2. Most tissues increases exponentially with the decrease in oxygen saturation of blood
   3. Tissues is dependent on a normal "vasomotion" of precapillary sphincters
   4. Tissues rendered temporarily ischemic will increase in proportion to the duration of the ischemia   Ref. pp. 271-273

SECTION I - THE CARDIOVASCULAR SYSTEM

137. LOCAL AUTOREGULATION OF BLOOD FLOW:
   1. Will permit a preservation of the same value of blood flow regardless of the value of perfusing pressure
   2. Will not allow the flow to decrease even if the perfusing pressure decreases below 10 mm Hg
   3. Has been found to exist only in the kidney
   4. Is more efficient when the changes are produced over a long time than when pressure changes occur acutely
   Ref. p. 273

138. LONG TERM ADJUSTMENT OF BLOOD FLOW TO <u>LOCAL</u> TISSUE NEEDS TAKES PLACE FUNDAMENTALLY THROUGH A PROCESS THAT INVOLVES MAINLY:
   1. Chronic hypertension in hypoxic tissue arterioles
   2. Decreased viscosity of blood when there is localized hypoxia
   3. Increased cardiac output when there is localized hypoxia
   4. Growth of new vessels in hypoxic tissue
   Ref. p. 274

139. THE MOST IMPORTANT FACTOR SEEMINGLY <u>RESPONSIBLE</u> FOR THE MAINTENANCE OF LONG TERM <u>LOCAL</u> REGULATION OF BLOOD FLOW THROUGH A GIVEN TISSUE IS THE MAINTENANCE OF:
   1. Blood pressure in the whole body
   2. Temperature of the local area
   3. A high oxygen saturation in the venous blood
   4. A normal tissue metabolism (locally)
   Ref. p. 276

140. IT IS TRUE TO SAY THAT:
   1. The heart muscle pumps forward variable volumes of blood returning to it even without nervous control
   2. An increase in extracellular fluid volume will increase the blood volume and the "mean systemic" pressure
   3. An increase in arterial pressure will eventually cause an increased excretion of water and salt
   4. A decrease in "mean systemic" pressure will reduce venous return of blood to the heart
   Ref. pp. 276-283

141. THE VASOMOTOR CENTER:
   1. Is located in the upper portions of the pons
   2. Fires tonically at a rate of 0.5-2 impulses/second
   3. May not be blocked by spinal anesthesia
   4. Has both excitatory and inhibitory components (in terms of blood pressure)
   Ref. p. 286

142. BLOOD PRESSURE IS INFLUENCED BY SEVERAL AREAS OF THE CENTRAL NERVOUS SYSTEM AMONG WHICH A POTENTIAL ROLE IS PLAYED BY THE:
   1. Hypothalamus
   2. Motor cortex
   3. Cingulate gyrus
   4. Cerebellum
   Ref. p. 287

SECTION I - THE CARDIOVASCULAR SYSTEM

143. DURING MUSCULAR EXERCISE VASOCONSTRICTING INFLUENCES MAY REACH THE SPINAL CORD <u>DIRECTLY</u> FROM (WITHOUT PASSING THROUGH THE VASOMOTOR CENTER) THE:
    1. Cerebellum
    2. Lower part of the medulla
    3. Anterior portion of the pons
    4. Motor cortex   Ref. p. 287

144. VASOCONSTRICTING NEUROTRANSMITTER SUBSTANCE(S) RELEASED AT THE ENDINGS OF VASOCONSTRICTOR NERVES IS(ARE):
    1. Acetylcholine
    2. Epinephrine
    3. DOPA
    4. Norepinephrine   Ref. p. 287

145. VASODILATING NEUROTRANSMITTER SUBSTANCE(S) RELEASED AT THE ENDS OF VASODILATOR NERVES IS(ARE):
    1. DOPA
    2. Epinephrine
    3. Norepinephrine
    4. Acetylcholine   Ref. p. 287

146. WHEN THE VASOMOTOR CENTER SHARPLY INCREASES ITS ACTIVITY:
    1. There is massive vasoconstriction in many territories
    2. The heart is stimulated to increase its output
    3. The veins are constricted and their capacity to store blood is reduced
    4. The carotid sinus reflex is depressed
                                                      Ref. p. 288

147. EMOTIONAL FAINTING IS USUALLY ASSOCIATED WITH:
    1. Vasodilation in skeletal muscles
    2. Very intense activity of the vasomotor center in the lower pons which causes massive vasoconstriction
    3. Activation of cholinergic fibers originated in the spinal cord
    4. A cerebellar inhibition of muscle tone
                                                      Ref. p. 288

148. ACUTE ISCHEMIA OF THE CENTRAL NERVOUS SYSTEM RESULTS IN A HYPERTENSIVE RESPONSE (RISE IN ARTERIAL BLOOD PRESSURE) WHICH IS PRESUMABLY DUE TO:
    1. Low pressure of the cerebrospinal fluid
    2. Consciousness of the hypoxia with a necessary participation of the cerebral cortex
    3. Ischemia of the carotid chemoreceptors
    4. Relative accumulation of carbon dioxide in the area of the vasomotor center   Ref. p. 289

149. A SUDDEN RISE IN PRESSURE OF THE CEREBROSPINAL FLUID TO EQUAL THE VALUE OF BLOOD PRESSURE WILL USUALLY GIVE RISE TO:
    1. Tachycardia
    2. An immediate increase in brain perfusion
    3. A complete heart block
    4. Hypertension   Ref. p. 289

SECTION I - THE CARDIOVASCULAR SYSTEM   35

150. ASCENDING IMPULSES TO THE CENTRAL NERVOUS SYSTEM THAT WILL CONTRIBUTE TO THE SO-CALLED BARORECEPTOR REFLEX ARE CARRIED BY FIBERS WITHIN THE:
1. Glossopharyngeal nerve
2. Hering's nerve
3. Vagus nerve
4. Hypoglossal nerve   Ref. p. 290

151. THE RANGE(S) OF ARTERIAL PRESSURE VALUES WHICH CAN ELICIT FIRING OF FIBERS ORIGINATED IN THE CAROTID SINUS NERVES LIES BETWEEN:
1. 0-50 mm Hg
2. 50-200 mm Hg
3. 200-250 mm Hg
4. 80-100 mm Hg   Ref. p. 290

152. THE FREQUENCY OF IMPULSES RECORDED IN A CAROTID SINUS NERVE ASCENDING TO THE CENTRAL NERVOUS SYSTEM:
1. Increases with diastole
2. Decreases with diastole
3. Decreases with systole
4. Increases with systole   Ref. p. 291

153. AN INCREASE IN THE RATE OF FIRING OF BARORECEPTOR IMPULSES:
1. Inhibits the vasomotor center
2. Excites the vagal center
3. Induces vasodilatation throughout the peripheral circulatory system
4. Decreases the rate of cardiac contractions
                                     Ref. p. 291

154. IN THE NORMAL PERSON, THE CHANGE IN POSITION FROM LYING DOWN IN BED TO STANDING ERECT IS ASSOCIATED WITH A(N):
1. Increased activity of the vasomotor center
2. Increased sympathetic tone
3. Decrease in vagal firing to the sino-atrial node
4. Maintenance of the mean arterial pressure
                                     Ref. p. 292

155. IN CHRONIC HYPERTENSION, THE RATE OF FIRING FROM BARO-RECEPTORS TO THE CENTRAL NERVOUS SYSTEM IS LIKELY TO:
1. Persist at a high level at the end of several weeks of high blood pressure
2. Decrease after a short initial rise
3. Regulate blood pressure effectively
4. Undergo adaptation   Ref. p. 292

156. WHEN THE CAROTID AND AORTIC CHEMORECEPTORS ARE PERFUSED WITH HYPOXEMIC BLOOD, THE VASOMOTOR CENTER WILL:
1. Respond by causing wide fluctuations of blood pressure even when the pressure is normal
2. Not respond at all
3. Be depressed
4. Be activated to a small extent and mainly if blood pressure is low
                                     Ref. p. 292

SECTION I - THE CARDIOVASCULAR SYSTEM

157. INCREASED WARMTH OF THE SKIN CAN CAUSE A REFLEX VASODILATION OF THE INVOLVED AREAS OF THE SKIN. PORTIONS OF THE CENTRAL NERVOUS SYSTEM NOT REQUIRED FOR THIS REFLEX (CAN BE EXPERIMENTALLY REMOVED WITHOUT SUPPRESSING THE REFLEX BY THEIR ABSENCE) ARE THE:
    1. Thalamus
    2. Pons
    3. Vasomotor center
    4. Spinal cord Ref. p. 293

158. THE MOST SENSITIVE VASCULAR AREAS CAPABLE OF RESPONDING TO THE WEAKEST VASOCONSTRICTING IMPULSES ORIGINATING IN THE CENTRAL NERVOUS SYSTEM (SYMPATHETIC) IS(ARE):
    1. Capillaries
    2. Lymphatics
    3. Arteries and arterioles
    4. Veins and venules Ref. p. 293

159. THE CONSTRICTION OF VEINS CAUSES MOST SIGNIFICANTLY A(N):
    1. Increase in peripheral resistance
    2. Diminution of blood flow returning to the heart
    3. Pooling of blood in the capillaries
    4. Decrease in their capacity Ref. p. 293

SELECT THE ONE MOST APPROPRIATE ANSWER:

160. VENOUS CONSTRICTION AS INDUCED BY INCREASED SYMPATHETIC ACTIVITY CAUSES A(N):
    A. Slower return of venous blood to the heart
    B. Increased cardiac output
    C. Increased peripheral resistance
    D. Longer "circulatory time" (time required for circulation of blood from veins in arm to lungs or tongue)
       Ref. p. 293

161. IN THE PRODUCTION OF SO-CALLED "TRAUBE-HERING WAVES" A MAJOR ROLE IS PLAYED BY:
    A. Consciousness
    B. The reticular activating system of the brain stem
    C. Visual stimuli
    D. The carotid baroreceptors
    E. The celiac plexus Ref. p. 294

162. RESPIRATORY MOVEMENTS ARE ASSOCIATED WITH (AND PARTLY CAUSE) FLUCTUATIONS IN BLOOD PRESSURE. THE HIGHEST RISE OF SYSTOLIC PRESSURE OCCURS AT THE TIME OF THE:
    A. Pause between inspiration and expiration
    B. End of expiration
    C. Beginning of inspiration
    D. End of inspiration Ref. p. 294

SECTION I - THE CARDIOVASCULAR SYSTEM   37

163. IT IS TRUE TO SAY THAT:
   A. During the beginning of inspiration the increased negativity of the thoracic pressure slows the return of pulmonary blood to the left heart cavities
   B. At the end of inspiration there is increased return of blood to the left heart cavities
   C. At the end of expiration the higher values of thoracic pressure (more positive) diminish the venous return to the right heart
   D. At the end of expiration there is a slight diminution of the cardiac output
   E. All of the above occur and none of the above phenomena occur with the exclusion of the others   Ref. p. 294

164. THE SECRETION OF ALDOSTERONE IS INCREASED BY:
   A. A decrease in extracellular fluid volume
   B. An increase in sodium concentration in plasma
   C. An increase in the cardiac output
   D. An increase in blood volume
   E. All of the above   Ref. p. 295

165. EPINEPHRINE, AS OPPOSED TO NOREPINEPHRINE, CAUSES:
   A. Vasodilation of cutaneous vessels
   B. Vasodilation of renal vessels
   C. Vasodilation of vessels in skeletal muscles
   D. Pupillary dilatation   Ref. p. 295

166. ACTIVATED KALLIKREIN ACTS ON:
   A. Alpha-2 globulin to release kallidin
   B. Albumin to release bradykinin
   C. Gamma-globulin to release carboxypeptidase
   D. Red blood cells causing their agglutination
                                             Ref. p. 295

167. ANGIOTENSIN IS:
   A. "Activated" by the presence of renin in circulation
   B. A powerful vasoconstrictor of arteries
   C. Formed more rapidly in conditions associated with hypotension
   D. Not a very effective constricting agent of veins
   E. All of the above   Ref. p. 295

168. THE FUNCTION OF VASOPRESSIN IS MAINLY TO CAUSE:
   A. An increase in blood pressure whenever there is a hypotensive tendency
   B. Constriction of veins
   C. Renal vasoconstriction
   D. Antidiuresis
   E. An increased secretion of aldosterone
                                             Ref. p. 295

169. KOROTKOW SOUNDS:
   A. Coincide with the turbulent eddies caused by the constriction of an artery
   B. Frequently appear at values of blood pressure above systolic
   C. Do not disappear below diastolic pressure
   D. Have an unchanging quality and intensity while the pressure in the cuff is gradually decreased   Ref. p. 299

SECTION I - THE CARDIOVASCULAR SYSTEM

170. THE RANGE OF ARTERIAL PRESSURES WITHIN WHICH THE CENTRAL NERVOUS SYSTEM ISCHEMIC RESPONSE MAY OCCUR IS SAID TO BE BETWEEN:
    A. 100-150 mm Hg
    B. 80-100 mm Hg
    C. 50-80 mm Hg
    D. 15-50 mm Hg   Ref. p. 300

171. SHORTLY AFTER THE DEVELOPMENT OF A SUDDEN ELEVATION OF MEAN ARTERIAL BLOOD PRESSURE, COMPENSATORY PHENOMENA INCLUDE:
    A. A capillary shift of intravascular fluid into the interstitial fluid
    B. A gradual "stress-relaxation" of distensible vessels
    C. Bradycardia through the carotid sinus reflex mechanism
    D. A slower rate of firing from the vasomotor center
    E. All of the above   Ref. p. 300, 301

172. IF AN ARTERIO-VENOUS FISTULA, WHICH IS RESPONSIBLE FOR A LOW PERIPHERAL RESISTANCE, IS SURGICALLY CLOSED, ARTERIAL BLOOD PRESSURE POST-OPERATIVELY WILL:
    A. Be permanently elevated
    B. Be lowered
    C. Be high at first but will be lowered after a few days
    D. Cause a final increase in extracellular fluid volume and blood volume
                                      Ref. p. 304

173. ALLOWING ANIMALS TO DRINK ISOTONIC SALINE INSTEAD OF WATER GIVES RISE TO ARTERIAL HYPERTENSION WHEN THE ANIMALS HAVE LOST ABOUT 75% OF THEIR KIDNEY FUNCTION. THE PROBABLE SEQUENCE OF EVENTS IN THIS CASE IS:
    A. Saline ingestion→Increased ECF→Increased cardiac output (CO)→Increased tissue perfusion→Increased peripheral resistance (PR)→Increased blood pressure
    B. Saline ingestion→Increased ECF→Increased PR→Increased blood pressure→Increased CO→Increased tissue perfusion
    C. Saline ingestion→Increased ECF→Increased peripheral resistance→Decreased tissue perfusion→Decreased cardiac output→Increased blood pressure
                                      Ref. p. 305

174. THE SO-CALLED "GOLDBLATT" HYPERTENSION CAUSED BY UNILATERAL KIDNEY DISEASE IS PROBABLY DUE TO THE:
    A. Inability of the remaining healthy kidney to excrete the excess of salt and water circulating in the body
    B. Spastic influence of renin secreted by the ischemic kidney
    C. Retention of water and salt due to an increased secretion of aldosterone
    D. Increased activity of the hypothalamus that responds to increased amounts of vasopressin   Ref. p. 307

175. IT IS TRUE TO SAY THAT IN CASES OF COARCTATION OF THE AORTA:
    A. The arterial pressures in the upper and lower extremities are equal
    B. The vascular resistance is low in the upper extremities
    C. There is a generalized vasoconstrictor influence acting on the circulation
    D. The blood flow through tissues of both upper and lower portions of the body is approximately normal
                                      Ref. p. 307

SECTION I - THE CARDIOVASCULAR SYSTEM

176. A PATIENT WHO HAS A PHEOCHROMOCYTOMA IN THE MEDULLA OF ONE OF THE ADRENAL GLANDS:
    A. Is subject to bouts of hypotension
    B. Secretes greater amounts of epinephrine in conjunction with sympathetic discharges (emotional stresses)
    C. Will excrete very small amounts of metabolites of epinephrine in the urine
    D. Cannot have its epinephrine properly blocked by pharmacologic agents
    Ref. p. 308

177. A PERSON WHOSE CARDIAC OUTPUT EQUALS 6.0 LITERS/MINUTE AND WHOSE BODY SURFACE AREA EQUALS 2.0 METERS^2 HAS A CARDIAC INDEX EQUAL TO:
    A. 2.0
    B. 3.0
    C. 4.0
    D. 5.0
    E. 6.0   Ref. p. 311

178. CARDIAC INDEX:
    A. In a given individual is essentially the same at age 25 and at age 50, provided the person retains the same surface area at both ages
    B. Rises to the same extent as the rise in oxygen consumption caused by increasing levels of exercise
    C. Decreases with an increase in basal metabolism
    D. Increases in the standing position if there is no motion at all
    Ref. p. 311

179. THE MAGNITUDE OF THE CARDIAC OUTPUT IS BASICALLY DETERMINED BY THE:
    A. Rate of cardiac contraction
    B. Force of cardiac contraction
    C. Coronary circulation
    D. Demands of the peripheral circulatory system
    Ref. p. 312

180. THE MOST IMPORTANT VARIABLE WHICH INCREASES CARDIAC OUTPUT DURING EXERCISE IS THE:
    A. Mental anticipatory stimulation of sympathetic discharge
    B. Compression of abdominal muscles which reduces the vascular bed by a purely mechanical effect and raises the mean systemic pressure
    C. Increased myocardial contractility due to increased circulating epinephrine
    D. Increased metabolism of muscle which increases blood flow through its dilated vessels   Ref. pp. 314-315

181. PATIENTS SUFFERING FROM VARIOUS DISEASES WHICH DIMINISH THE LEVEL OF PERIPHERAL RESISTANCE (BERI-BERI, PAGET'S DISEASE) WILL MOST LIKELY SHOW:
    A. A decreased cardiac output
    B. An increased cardiac output
    C. An unmodified cardiac output
    D. Anoxia in all of these cases   Ref. p. 315

SECTION I - THE CARDIOVASCULAR SYSTEM

182. AN INCREASED ACTIVITY OF THE SYMPATHETIC SYSTEM AT THE LEVEL OF THE HEART WITHOUT A CONCOMITANT INCREASE IN THE PERIPHERAL SYMPATHETIC TONE WILL:
    A. Increase significantly the cardiac output
    B. Decrease significantly the cardiac output
    C. Not alter the cardiac output to any significant degree
    D. Decrease the "permissive" level of pumping blood
    Ref. p. 312, 315

183. REMEMBERING THAT THE HEART IS FOR THE MOST PART NOT THE MAIN DETERMINANT OF THE CARDIAC OUTPUT BUT THAT THE PERIPHERAL CIRCULATION IS THE KEY VARIABLE IN THE REGULATION OF SUCH CARDIAC OUTPUT, IT IS ALSO TRUE THAT CARDIAC:
    A. Damage (infarction) does not alter the potential cardiac output significantly
    B. Damage will nevertheless decrease the potential cardiac output significantly
    C. Damage can be completely compensated for by an increased venous return
    D. Damage can be completely compensated for by an increased sympathetic tone reaching the myocardium   Ref. pp. 315-316

184. AN INCREASED "EFFECTIVENESS" OF VENTRICULAR CONTRACTION (CARDIAC OUTPUT PER RIGHT ATRIAL PRESSURE) MAY INCREASE BY:
    A. Sympathetic stimulation
    B. Cardiac hypertrophy
    C. Decreased arterial pressure
    D. Inhibition of parasympathetic influence on the heart
    E. All of the above   Ref. p. 316

185. THE FUNCTIONAL CURVE THAT RESULTS FROM PLOTTING CARDIAC OUTPUT (ORDINATES) VERSUS RIGHT ATRIAL PRESSURE (ABSCISSAE) WILL, IF THE INTRAPLEURAL PRESSURE IS INCREASED FROM -4 TO -2 mm Hg:
    A. Have a much reduced slope
    B. Be shifted in parallel fashion to the right
    C. Be shifted in parallel fashion to the left
    D. Not shift at all   Ref. p. 316

186. THE ABOVE DESCRIBED CHANGES MEAN THAT DURING FORCED EXPIRATION THE VENTRICLES WILL HAVE:
    A. The same contractility as during inspiration
    B. A much increased contractility
    C. A much decreased contractility
    D. A mild decrease in contractility   Ref. p. 316

187. THE RATE OF RETURN OF BLOOD TO THE RIGHT ATRIUM ("VENOUS RETURN") IS CAUSED BY THE FACT THAT THERE IS ABOUT 7 mm Hg OF PRESSURE DIFFERENCE BETWEEN THE PERIPHERAL VEINS (TRULY MEAN SYSTEMIC PRESSURE) AND THE RIGHT ATRIUM. AN INCREASE IN RIGHT ATRIAL PRESSURE TO A LEVEL EQUAL TO 10 mm Hg WILL RESULT IN:
    A. Decreased venous return
    B. Cessation of venous return
    C. Decrease in ventricular contractility
    D. No change in venous return   Ref. p. 318

SECTION I - THE CARDIOVASCULAR SYSTEM

188. AN INCREASE IN THE NEGATIVITY OF THE THORACIC PRESSURE FROM -2 TO -10 mm Hg WILL CAUSE THE VENOUS RETURN TO BE:
    A. Significantly increased
    B. Unchanged
    C. Decreased
    D. At first greatly increased and then decreased
    Ref. p. 318

189. VENOUS RETURN TO THE HEART IS FACILITATED BY:
    A. The amount of blood filling the circulatory system
    B. The tone of the sympathetic system on arterioles and veins
    C. The negativity of the thoracic pressure
    D. The activity of skeletal muscles
    E. All of the above
    Ref. p. 318

190. CARDIAC OUTPUT MEASURED THROUGH VERY FREQUENT INTERVALS DURING SYSTOLE AND DIASTOLE REVEALS THAT DURING SYSTOLE:
    A. Output reaches a high value only late in systole
    B. Output has a negative value for a few milliseconds in the latter part of systole
    C. Output is essentially constant throughout systole due to a fixed capillary runoff
    D. A major fraction of ventricular output returns to the ventricle at the end of systole
    Ref. p. 321

191. ACCORDING TO THE FICK PRINCIPLE THE QUANTITY OF OXYGEN NEWLY ADDED TO BLOOD PASSING THROUGH PULMONARY ALVEOLI IN THE UNIT TIME IS PROPORTIONAL TO THE VOLUME OF BLOOD PERFUSING THE LUNGS IN THE SAME UNIT OF TIME. GIVEN AN OXYGEN CONSUMPTION EQUAL TO 200 ml O_2/MINUTE, A PULMONARY ARTERY OXYGEN CONTENT EQUAL TO 0.16 ml O_2/ml BLOOD, A LEFT ATRIAL OXYGEN CONTENT EQUAL TO 0.20 ml O_2/ml BLOOD, THE CARDIAC OUTPUT EQUALS:
    A. 1,000 ml blood/minute
    B. 2,000 ml blood/minute
    C. 3,000 ml blood/minute
    D. 4,000 ml blood/minute
    E. 5,000 ml blood/minute
    Ref. p. 322

192. IF 5 MILLIGRAMS OF T-1824 DYE ARE INJECTED AT TIME ZERO IN A PERIPHERAL VEIN, AND THE FIRST DYE APPEARS IN AN ARM ARTERY 5 SECONDS LATER, REACHING AN AVERAGE CONCENTRATION EQUAL TO 0.005 mg/ml, AND ESSENTIALLY DISAPPEARING (BY EXTRAPOLATION) FROM CIRCULATION 20 SECONDS AFTER THE INJECTION, WE CAN CALCULATE THE CARDIAC OUTPUT TO BE EQUAL TO:
    A. 1,000 ml blood/minute
    B. 2,000 ml blood/minute
    C. 3,000 ml blood/minute
    D. 4,000 ml blood/minute
    E. 5,000 ml blood/minute
    Ref. pp. 322-323

SECTION I - THE CARDIOVASCULAR SYSTEM

193. AN INVARIABLE PHYSIOLOGICAL PARAMETER PRESENT IN ALL CASES OF SHOCK IS A DECREASED:
   A. Blood pressure
   B. Cardiac output
   C. Alveolar ventilation
   D. Arterial oxygen saturation
   E. Ventilation pulmonary perfusion ratio
   Ref. p. 325

194. CARDIAC OUTPUT IS RELATED TO VENOUS RETURN AND MEAN SYSTEMIC PRESSURE AND BLOOD VOLUME IN A VERY SPECIFIC CAUSE-EFFECT RELATIONSHIP. WHEN A MODERATELY SEVERE HEMORRHAGE DIMINISHES THE BLOOD VOLUME TO A SUFFICIENT DEGREE, WE CAN ASSUME THAT THE FOLLOWING SEQUENCE LEADS TO SHOCK:
   A. Hemorrhage→Diminished blood volume (<B.V)→Diminished mean systemic pressure (<S.P.)→Diminished venous return (<V.R.)→ Diminished cardiac output (<C.O.)→Shock
   B. Hemorrhage→<B.V.→<C.O.→<V.R.→<S.P.→Shock
   C. Hemorrhage→<B.V.→<V.R.→<C.O.→<S.P.→Shock
   D. Hemorrhage→<B.V.→<S.P.→<C.O.→<V.R.→Shock
   Ref. p. 325

195. A PERSON WHO HAS A BLOOD VOLUME EQUAL TO 4,000 ml MAY LOSE WITHOUT DANGER OF SUFFERING HYPOVOLEMIC SHOCK A MAXIMUM OF:
   A. 300 ml blood (7.5% of the blood volume)
   B. 400 ml blood (10.0% of the blood volume)
   C. 800 ml blood (20% of the blood volume)
   D. 1200 ml blood (30% of the blood volume)
   E. 1800 ml blood (45% of the blood volume)
   Ref. pp. 325-326

196. SYMPATHETIC REFLEXES WHICH OCCUR FOLLOWING ACUTE BLOOD LOSS TEND TO MAINTAIN ABOVE ALL OTHER PHYSIOLOGICAL PARAMETERS THE ORIGINAL VALUE OF:
   A. Alveolar ventilation
   B. Cardiac output
   C. Blood pressure
   D. Venous return
   E. Blood volume
   Ref. p. 326

197. SYMPATHETIC REFLEXES OCCURRING SHORTLY AFTER THE ONSET OF AN ACUTE BLOOD LOSS RESTRICT THE CAPILLARY PERFUSION OF SEVERAL TISSUES. THE LEAST RESTRICTED OF ORGANS LISTED BELOW IS THE:
   A. Liver
   B. Skin
   C. Intestine
   D. Kidney
   E. Heart
   Ref. p. 326

SECTION I - THE CARDIOVASCULAR SYSTEM 43

198. FROM AN ANALYSIS OF VENTRICULAR FUNCTION CURVES (PLOTTING CARDIAC OUTPUT AS A FUNCTION OF RIGHT ATRIAL PRESSURE) IN SEVERE AND IRREVERSIBLE EXPERIMENTAL HYPOVOLEMIC SHOCK IN DOGS ONE MAY PREDICT THAT THE TIME REQUIRED FOR THE HEART TO LOSE 50% OF ITS CONTRACTILE CAPACITY (AFTER THE INITIAL BLOOD LOSS) IS APPROXIMATELY EQUAL TO:
    A. One hour
    B. Two hours
    C. Four hours
    D. Seven hours
    E. Ten hours   Ref. Fig. 28-4, p. 327

199. A DIMINISHED PERFUSION OF BLOOD THROUGH THE VASOMOTOR CENTER WILL CAUSE IRREVERSIBLE DAMAGE IN THESE AREAS OF THE CENTRAL NERVOUS SYSTEM. THE TIME REQUIRED FOR THE DAMAGE TO BECOME IRREVERSIBLE IS APPROXIMATELY EQUAL TO:
    A. One minute
    B. Three minutes
    C. Ten minutes
    D. Fifty minutes
    E. Thirty minutes   Ref. p. 328

200. THERE IS A GOOD CORRELATION BETWEEN OXYGEN DEFICIT AND PROGNOSIS OF SHOCK. A DEFICIT IN OXYGEN CONSUMPTION EQUAL TO 1 ml Kg minute WILL TEND TO CAUSE IRREVERSIBLE SHOCK IN APPROXIMATELY:
    A. Thirty minutes
    B. One hour
    C. Two hours
    D. Three hours
    E. Four hours   Ref. p. 330

201. THE PHARMACOLOGIC "SUPPORT" OF CARDIAC CONTRACTILITY (USING OUABAIN) HAS BEEN FOUND (IN DOGS) CAPABLE OF EXTENDING THE TIME REQUIRED, FOR AN OXYGEN DEFICIT OF 1 ml kg minute, TO CAUSE IRREVERSIBILITY, FROM THE VALUE REFERRED TO IN THE PREVIOUS QUESTION FOR AN ADDITIONAL:
    A. One hour
    B. Two hours
    C. Three hours
    D. Four hours
    E. Five hours   Ref. Fig. 28-6, p. 330

SECTION I - THE CARDIOVASCULAR SYSTEM

202. KNOWING THAT OUABAIN SPECIFICALLY STIMULATES MYOCARDIAL FIBER CONTRACTILITY ONE MAY CONCLUDE FROM THE EFFECTS INDICATED BY THE PREVIOUS EXPERIMENT THAT IN HYPOVOLEMIC SHOCK:
   A. Transfusions with oxygenated blood should restore circulatory function completely, even if administered at the end of the period of oxygen deficit
   B. Vasoconstriction caused by the infusion of sympathomimetic drugs should be effective if this infusion is sufficiently prolonged
   C. Irreversibility probably is originated in the primary deterioration of the splanchnic area and the kidney's parenchyma
   D. The hypoxic and damaged myocardium is a primary cause of irreversibility
   E. Ouabain could restore myocardial function completely if infused properly, regardless of the cummulative oxygen deficit
   Ref. p. 330

203. THE MOST LOGICAL FORM OF THERAPY OF SHOCK INDUCED BY SEVERE BURNS OF THE SKIN, AMONG THOSE LISTED BELOW, IS:
   A. Transfusion of whole blood
   B. Infusion of large volumes of hypertonic saline
   C. Infusion of large volumes of hypotonic dextrose
   D. Transfusion of packed red blood cells
   E. Transfusion of plasma   Ref. p. 331

204. THE CRITICAL DISTURBANCE LEADING TO CARDIOVASCULAR SHOCK WHEN THERE IS A TOTAL LOSS OF VASOMOTOR TONE CONSISTS OF:
   A. An impaired cardiac contractility
   B. A reduced vascular compliance
   C. A reduced oxygen consumption
   D. A reduced venous return
   E. A reduced capillary permeability   Ref. p. 331

205. THE SO-CALLED TRENDELENBURG POSITION IS A RECOMMENDED POSTURE FOR PATIENTS IN SHOCK BECAUSE IT TENDS TO FACILITATE:
   A. Capillary vasomotor contractility
   B. Cardiac contractility
   C. Venomotor contractility
   D. Venous return to the heart
   E. Cerebral arteriolar constriction   Ref. p. 331

206. A PERSON WHO HAS SUFFERED AN EXTENSIVE CONTUSION OF ONE OF THE EXTREMITIES CAUSED BY A LARGE BLUNT OBJECT, WITHOUT APPARENT HEMORRHAGE OR FRACTURE OF BONES, IS LIKELY TO HAVE SHOCK DUE TO:
   A. Endotoxin
   B. Loss of plasma volume
   C. Expansion of the extracellular fluid volume due to excessive secretion of aldosterone
   D. Excessive vasoconstriction of arterioles in the traumatized area
   E. Bacterial invasion of traumatized area by previously inapparent bacteremia   Ref. p. 333

SECTION I - THE CARDIOVASCULAR SYSTEM

207. AN IMPORTANT CLINICAL PARAMETER FOUND IN SHOCK, DUE TO AN OVERALL DEPRESSED TISSUE METABOLISM IS:
    A. Fever
    B. Hypoventilation
    C. Hypothermia
    D. Mental excitability   Ref. p. 333

208. RENAL INSUFFICIENCY APPEARING SEVERAL DAYS AFTER A HYPO-TENSIVE EPISODE (OPERATION, TRAUMA) IS MOST LIKELY DUE TO A:
    A. Permanent suppression of glomerular filtration
    B. Tubular necrosis
    C. Massive precipitation of crystalline matter in tubules and massive urinary obstruction
    D. Metabolic alkalosis   Ref. p. 333

209. SYMPATHOMIMETIC DRUGS INFUSED OVER PROLONGED PERIODS OF TIME, DURING HEMORRHAGIC SHOCK, ARE OF LITTLE FINAL VALUE BECAUSE:
    A. Further vasoconstriction leads to cerebral ischemia
    B. Adrenergic receptors respond paradoxically to exogenous amines during shock
    C. Circulatory reflexes are already maximally activated
    D. Coronary constriction causes angina and myocardial infarction
                                         Ref. p. 334

210. SYMPATHOLYTIC DRUGS SHOULD BE IDEALLY COMBINED IN THE THERAPY OF SHOCK WITH THE SIMULTANEOUS ADMINISTRATION OF:
    A. Hyperbaric oxygen
    B. Large volumes of fluid isotonic with the extracellular fluid
    C. Packed red blood cells
    D. Sympathomimetic drugs   Ref. p. 334

211. THE RATIONALE OF GIVING A SYMPATHOMIMETIC DRUG IN ANA-PHYLACTIC SHOCK IS BASED ON THE PROPOSED ACTION OF THESE DRUGS WHICH SEEM TO:
    A. Counteract the vasodilation induced by histamine
    B. Prevent further release of histamine
    C. Prevent antigen-antibody combination
    D. Prevent the binding of complement Ref. p. 334

212. IN THE MAJORITY OF FORMS OF CARDIAC FAILURE, KNOWN AS CONGESTIVE HEART FAILURE, TWO PARAMETERS VARY FROM THE NORMAL VALUES AT THE SAME TIME:
    A. The cardiac output increases while the atrial pressure and end diastolic ventricular pressure is lowered
    B. The cardiac output decreases while the atrial pressure and end diastolic ventricular pressure is increased
    C. The cardiac output increases while the atrial pressure and end diastolic ventricular pressure is increased
    D. The cardiac output decreases while the atrial pressure and end diastolic ventricular pressure is decreased
                                         Ref. p. 338

SECTION I - THE CARDIOVASCULAR SYSTEM

213. SYMPATHETIC REFLEXES TAKING PLACE WITHIN <u>SECONDS</u> OF AN ACUTE HEART FAILURE WILL CAUSE THE CARDIAC OUTPUT TO:
   A. Decrease and the right atrial pressure to be diminished
   B. Increase and the right atrial pressure and end diastolic ventricular pressure to be increased
   C. Decrease and the right atrial pressure and end diastolic ventricular pressure to be increased
   D. Increase and the right atrial pressure and end diastolic ventricular pressure to be decreased   Ref. p. 338

214. IN MODERATE CARDIAC FAILURE, FLUID RETAINED IN THE EXTRA-CELLULAR FLUID SPACE BY A) A LOWER CARDIAC OUTPUT, AND B) AN INCREASED ALDOSTERONE SECRETION, TENDS TO:
   A. Decrease cardiac contractility
   B. Stimulate further aldosterone secretion
   C. Increase venous return to the heart
   D. Decrease atrial pressure   Ref. p. 339

THE FOLLOWING CURVES ARE OBTAINED BY PLOTTING CARDIAC OUTPUT (AND VENOUS RETURN) AS A FUNCTION OF RIGHT ATRIAL PRESSURE. FOUR PHYSIOLOGICAL CONDITIONS ARE SHOWN:
   1. A normal condition
   2. An acute cardiac decompensation
   3. An early compensatory stage obtained by sympathetic discharge
   4. A late compensatory stage obtained by fluid retention

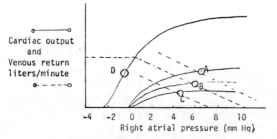

Redrawn from Fig.29-4 in Guyton's Textbook of Medical Physiology, 4th ed. W.B. Saunders Co.

THERE IS AN ASSOCIATION BETWEEN THE FOLLOWING CONDITIONS AND THE CORRESPONDING POINTS IN THE GRAPH:

215. ____ The normal condition.
216. ____ The acute cardiac decompensation
217. ____ Initial compensation due to sympathetic discharge
218. ____ Late compensation due to accumulation of fluid
                                        Ref. Fig. 29-4, p. 340

SECTION I - THE CARDIOVASCULAR SYSTEM

219. IN CARDIAC FAILURE, THE RETENTION OF FLUID, WHICH AT FIRST IS A COMPENSATORY FACTOR, BECOMES, WHEN EXCESSIVE IN MAGNITUDE, A DELETERIOUS FACTOR BECAUSE:
    A. The kidney retains with the excess of water too many products which should be excreted
    B. The osmolarity of extracellular water is abnormally high due to excessive sodium retention
    C. The contractility of the heart muscle cannot cope with this increase in blood volume and degree of distension
    D. An acid-base imbalance is produced, due to the high sodium bicarbonate concentration Ref. p. 341

220. IF CIRCULATORY FAILURE OCCURS RAPIDLY DUE TO AN ACUTE CARDIAC FAILURE (MASSIVE MYOCARDIAL INFARCTION) THE SYSTEMIC INTRACAPILLARY PRESSURE WILL:
    A. Rise from a mean value of 25 mm Hg
    B. Fall from a mean value of 25 mm Hg
    C. Undergo no change
    D. Pulsate in cycles synchronous with the cardiac systole and diastole
       Ref. p. 342

221. AS A CONSEQUENCE OF THE CHANGES DESCRIBED IN THE PREVIOUS QUESTION IT IS MOST LIKELY THAT EDEMA WILL:
    A. Occur rapidly (within a few minutes) after the initiation of cardiac failure
    B. Not occur until the renal and endocrine system cause the expansion of the extracellular fluid (within several hours or days)
    C. Occur mainly due to an increased capillary pressure secondary to a concomitant loss of arteriolar tone
    D. Not occur if sodium ingestion is not in excess of 5 gms sodium chloride per day Ref. p. 342

222. THE AREA OF LOCAL CIRCULATION WHICH REQUIRES IMMEDIATE RESTORATION IF IRREVERSIBLE CHANGES ARE TO BE PREVENTED IN CARDIOGENIC SHOCK IS THAT WHICH PERFUSES THE:
    A. Renal cortex
    B. Renal medulla
    C. Intestinal mucosa
    D. Liver
    E. Myocardium Ref. p. 343

223. IN THE CASE OF ACUTE LEFT VENTRICULAR FAILURE THE TRANSUDATION OF EXCESS OF FLUID IN THE ALVEOLAR SPACE FROM PULMONARY CAPILLARIES REQUIRES THAT THE INTRACAPILLARY PRESSURE EXCEED THE ONCOTIC PRESSURE BY A MINIMUM OF:
    A. 10 mm Hg
    B. 5 mm Hg
    C. 2 mm Hg
    D. 1 mm Hg Ref. p. 343

SECTION I - THE CARDIOVASCULAR SYSTEM

224. IF IN A PATIENT WITH LEFT VENTRICULAR FAILURE THE PULMONARY CAPILLARY PRESSURE RISES TO A HIGHER LEVEL THAN NORMAL (8 mm Hg) BUT BELOW THE LEVEL OF ONCOTIC PRESSURE (25 mm Hg), THE CHANCES THAT THE PATIENT WILL SUFFER A PULMONARY EDEMA ARE:
   A. Non-existent
   B. High, due to fluid retention within few days
   C. Very slight, because of the excellent defense provided by the osmotic forces provided by plasma proteins
   D. Low, due to simultaneous loss of contractility of the right ventricle
   Ref. p. 344

225. A CONSTANT FEATURE OF SO-CALLED "HIGH OUTPUT CARDIAC FAILURE" IS A(N):
   A. Weak contractility of the left ventricle
   B. Increased venous pressure of the pulmonary circuit
   C. Increased volume of venous return to the heart
   D. Systemic arteriolar vasoconstriction with low capillary pressure in the systemic circuit
   Ref. p. 345

226. THE MAXIMAL INTENSITY OF PHYSICAL ENERGY CAUSED BY VIBRATIONS GENERATED BY THE CARDIAC CYCLE:
   A. Lies below the lower limit of auditory perception (40 cycles per second)
   B. Is ideally placed where the threshold for auditory sensation is minimal
   C. Equals 2000 cycles per second
   D. Lies above the range of frequency where the spoken voice is located (256-2000 cps)
   Ref. p. 349

227. THE LOUDNESS (INTENSITY OF PHYSICAL ENERGY) OF THE CARDIAC SOUNDS IS DIRECTLY PROPORTIONAL TO THE:
   A. Length of the valves (from insertion to edge of cusps)
   B. Mobility of the cusps
   C. Rigidity of the cusps
   D. Rate of change of pressure differences across the closing orifices
   Ref. p. 349

228. THE THIRD HEART SOUND OCCURS:
   A. Immediately after the second heart sound
   B. In the first third of ventricular diastole
   C. In the middle third of ventricular diastole
   D. In the last third of ventricular diastole
   Ref. p. 350

229. THE MURMUR CAUSED BY AORTIC STENOSIS IS DUE TO:
   A. Vibrations caused by the very forceful left ventricular contraction
   B. Turbulent eddies caused by a high speed jet of blood in the aorta
   C. The sudden distension of the aortic walls suddenly distended by a high ventricular systolic stroke volume
   D. The sudden distension of the rigid aortic valves
   Ref. p. 351

SECTION I - THE CARDIOVASCULAR SYSTEM   49

230. LOUD AORTIC MURMURS HEARD AT THE LAST THIRD OF DIASTOLE
      AS WELL AS DURING THE EARLY AND MIDDLE THIRDS OF DIASTOLE
      SUGGEST:
      A. Less severe degrees of aortic insufficiency
      B. Very severe aortic insufficiency
      C. Calcified aortic valves
      D. Pulsatile flow of blood strictly limited to the aorta
                                                        Ref. pp. 351-352

231. THE FLOW OF REGURGITANT BLOOD GOING FROM LEFT VENTRICLE
      INTO THE LEFT ATRIUM IN CASES OF MITRAL VALVE INSUFFICIENCY
      TAKES PLACE FROM THE ANTERIOR PORTION OF THE CHEST
      CAVITY TO:
      A. A more posteriorly placed area of the chest cavity
      B. A more caudally placed area of the chest cavity
      C. The right side of the chest cavity
      D. The left side of the chest cavity   Ref. p. 352

232. MITRAL INSUFFICIENCY CAUSES A MURMUR THAT COINCIDES WITH
      THE:
      A. Protodiastole of the left ventricle
      B. Left atrial contraction
      C. Left ventricular contraction
      D. Left ventricular relaxation   Ref. p. 352

233. MITRAL STENOSIS CAUSES A MURMUR WHICH IS:
      A. High pitched, pansystolic, loud
      B. Low pitched, mid-diastolic, soft
      C. Protodiastolic, loud
      D. Holosystolic, ejection type, almost inaudible
                                                   Ref. p. 352

234. IN AORTIC INSUFFICIENCY OF SEVERE DEGREE A COMBINATION OF
      PHYSIOLOGICAL ABNORMALITIES LEAD TO:
      A. High end-diastolic ventricular volume (EDVV)
         Low end-diastolic aortic pressure (EDAP)
         Low cardiac output (CO)
         Large systolic ventricular stroke volume (SVSV)
      B. Low EDVV
         High EDAP
         High CO
         Small SVSV
      C. Low EDVV
         High EDAP
         Low CO
         Large SVSV
      D. High EDVV
         Low EDAP
         High CO
         Small SVSV   Ref. p. 353

SECTION I - THE CARDIOVASCULAR SYSTEM

235. THE MOST VALUABLE DIAGNOSTIC PHYSIOLOGICAL FEATURE THAT CHARACTERIZES AORTIC STENOSIS IS A:
A. Weak aortic second sound
B. Systolic murmur over the aortic region
C. Relatively large systolic pressure gradient between left ventricle and the aorta
D. Systolic "click"   Ref. p. 353

236. THE EFFECTIVENESS OF CARDIAC CONTRACTION IN SEVERE MITRAL DISEASES ASSOCIATED WITH A DILATED LEFT ATRIUM MAY BE DIMINISHED BY THE FREQUENT OCCURRENCE OF:
A. A shorter P-R interval than in the normal person
B. A longer P-R interval than in the normal person
C. Atrial fibrillation
D. Second degree heart block with Wenckebach phenomenon
                                         Ref. p. 353

237. A CARDIAC CONGENITAL ANOMALY WHICH ALLOWS PART OF THE LEFT VENTRICULAR OUTPUT TO BE SHUNTED TO THE RIGHT VENTRICLE WILL CAUSE:
A. No overloading of the left ventricle
B. Eventual failure of the left ventricle as well as the right ventricle
C. Part of the blood reaching the brain to have a diminished oxygen saturation
D. A very low end-diastolic volume of the left ventricle
                                         Ref. p. 354

238. IN THE FETUS THE PRESSURE IN THE PULMONARY ARTERY IS:
A. Lower than in the aorta
B. Higher than in the aorta
C. Equal than in the aorta
D. Almost zero   Ref. p. 354

239. IN PATIENTS WITH PERSISTENT DUCTUS ARTERIOSUS, LEFT VENTRICULAR OUTPUT IS:
A. Not elevated
B. Only partially oxygenated
C. Two to five times greater than right ventricular output
D. One to three times less than right ventricular output
                                         Ref. p. 355

240. AS A CONSEQUENCE OF THE ALTERED PHYSIOLOGICAL CONDITIONS REFERRED TO IN THE PREVIOUS QUESTION, THE PATIENT WITH PERSISTENT DUCTUS ARTERIOSUS:
A. Tends to be cyanotic even at rest
B. Has a low systolic pulmonary arterial pressure
C. Has a very low cardiac reserve
D. Has a low aortic pulse pressure   Ref. p. 355

241. IN THE PATIENT WITH PERSISTENT DUCTUS ARTERIOSUS, THE RIGHT VENTRICLE TENDS TO UNDERGO HYPERTROPHY BEACUSE OF:
A. Persistent cyanosis
B. The blood that shunts from the pulmonary artery to the aorta
C. The increased volume of blood flow through the pulmonary circuit which increases the pressures in the pulmonary artery
D. The arteriolar dilatation of the pulmonary arterioles
                                         Ref. p. 355

SECTION I - THE CARDIOVASCULAR SYSTEM

242. A HIGHLY DIAGNOSTIC FINDING IN A PATIENT AFFLICTED BY AN INTERVENTRICULAR SEPTAL DEFECT IS A:
    A. Higher left ventricular systolic pressure than the right ventricular systolic pressure
    B. Sudden increase in oxygen content in blood sampled in the left ventricle
    C. Sudden increase in oxygen content in blood sampled from the right ventricle
    D. Blowing diastolic murmur best heard over the midsternal region
    Ref. p. 356

243. IN THE MAJORITY OF HUMANS THE PRESSURE WITHIN THE LEFT ATRIUM IS:
    A. Equal to that of the right atrium
    B. Greater than that of the right atrium
    C. Lower than that of the right atrium
    D. Fluctuating more widely with respiratory movements than similar pressures within the right atrium   Ref. p. 356

244. DUE TO REASONS INDICATED BY THE LAST QUESTION, THE PATIENT WHO HAS A LARGE DEFECT IN THE SEPTUM THAT SEPARATES THE LEFT ATRIUM FROM THE RIGHT ATRIUM HAS:
    A. Blood shunted from left atrium to right atrium
    B. Blood shunted from right atrium to left atrium
    C. No blood shunted at all
    D. Partially deoxygenated blood reaching systemic vessels
    Ref. p. 356

245. IN THE SO-CALLED TETRALOGY OF FALLOT THERE IS, AMONG OTHER ABNORMALITIES, A(N):
    A. Atrial septal defect
    B. Stenosis of the pulmonary artery
    C. Marked dilatation of the pulmonary artery which overrides the septum between ventricles
    D. Normal partial pressure of oxygen in arterial blood
    Ref. p. 356

246. A SURGICAL PROCEDURE WHICH BENEFITS PATIENTS WITH TETRALOGY OF FALLOT INCLUDES THE IATROGENIC DEVELOPMENT OF A(N):
    A. Inter-atrial septal defect
    B. Vascular communication between a systemic artery and the pulmonary artery
    C. Vascular communication between a systemic artery and a pulmonary vein
    D. Vascular communication between a pulmonary vein and the right atrium
    Ref. p. 357

247. THE LEAST DEGREE OF HYPERTROPHY OF THE LEFT VENTRICLE OCCURS IN PATIENTS WITH:
    A. Aortic insufficiency
    B. Aortic stenosis
    C. Mitral insufficiency
    D. Patent ductus arteriosus
    E. Mitral stenosis   Ref. p. 357

SECTION I - THE CARDIOVASCULAR SYSTEM

248. THE RIGHT VENTRICLE MAY BE EXPECTED TO HAVE A GREATER THICKNESS THAN THAT OF THE LEFT VENTRICLE IN CASES OF:
    A. Aortic stenosis
    B. Aortic insufficiency
    C. Patent ductus arteriosus
    D. Tetralogy of Fallot
    E. All of the above   Ref. p. 358

249. WHEN A NORMAL HUMAN BEING PERFORMS PHYSICAL EXERCISE THE TOTAL CARDIAC OUTPUT (CO) AND THE CORONARY CIRCULATION (CC) INCREASE TO MEET THE DEMANDS OF THE EXCESS OF WORK IMPOSED ON THE WHOLE BODY AND ON THE HEART. THIS INCREASE OCCURS SO THAT THE RATIO CC/CO EQUALS:
    A. One
    B. Less than one
    C. More than one   Ref. p. 359

250. VARIOUS FACTORS ACT TO MODIFY THE RATE OF FLOW OF BLOOD THROUGH CAPILLARIES OF THE CORONARY CIRCULATION AT DIFFERENT TIMES OF THE CARDIAC CYCLE. IT HAS BEEN ESTIMATED THAT THE MAXIMAL RATE OF FLOW OCCURS DURING THE:
    A. Early part of systole
    B. Late part of systole
    C. Early part of diastole
    D. End of diastole   Ref. Fig. 31-1, p. 359

251. THE INCREASE IN BLOOD FLOWING THROUGH CORONARY VESSELS DURING EXERCISE IS LARGELY DUE TO A REGULATION WHICH INVOLVES A(N):
    A. Parasympathetic mediated vasodilation
    B. Adrenergic relaxation of the tension exerted by myocardial fibers
    C. Increase in coronary pressure due to a concomitant higher aortic pressure
    D. Oxygen dependent mechanism which probably determines the release of vasodilator substances during brief periods of tissue hypoxia
                                           Ref. p. 361

252. THE WORK OF THE HEART IS DIRECTLY PROPORTIONAL TO:
    A. The arterial pressure and the cardiac output
    B. The velocity of blood flow through the aorta and the reciprocal of the peripheral resistance
    C. Vagal impulses reaching the heart and the magnitude of the resting membrane potential at the sino-atrial node
    D. The time between cardiac cycles and the arterio-venous oxygen difference between coronary arteries and coronary sinus
                                           Ref. p. 361

253. WHEN THE HEART BECOMES DILATED, THE LAW OF LAPLACE ALLOWS THE CONCLUSION THAT CARDIAC TENSION FOR A GIVEN INCREMENT IN SYSTOLIC PRESSURE WILL UNDER THE NEW CIRCUMSTANCES:
    A. Be smaller than before dilation
    B. Be greater than before dilation
    C. Rise in inverse proportion to the increase in radius
    D. Decrease in inverse proportion to the increase in pressure to be obtained during systole   Ref. p. 361

SECTION I - THE CARDIOVASCULAR SYSTEM

254. VENTRICULAR FIBRILLATION DEVELOPS WITH RELATIVELY HIGH FREQUENCY IN A HEART WHICH HAS RECENTLY UNDERGONE A PARTIAL INFARCTION, BECAUSE THERE IS:
    A. Leakage of cellular potassium from the dead tissue and a local higher concentration of potassium ions in myocardial extracellular fluid
    B. A shorter refractory period in infarcted muscle
    C. Hypersensitivity of infarcted tissue to normal levels of epinephrine
    D. A very high resting membrane potential developed in the peri-infarcted areas   Ref. p. 365

255. THE PRINCIPAL FACTOR REGULATING THE RATE OF BLOOD FLOW TO THE BRAIN IS THE CONCENTRATION IN CEREBRAL TISSUES OF:
    A. Oxygen
    B. Carbon dioxide
    C. Hydrogen ions
    D. Lactate ions
    E. Pyruvic ions   Ref. p. 367

256. THE LEVEL OF NEURONAL ACTIVITY IS:
    A. Increased by increased levels of carbon dioxide
    B. Decreased by increased levels of carbon dioxide
    C. Not significantly affected by increased levels of carbon dioxide
    D. Far more affected by variations in lactate and other fixed acids than by carbon dioxide variations   Ref. p. 368

257. A DECREASE IN OXYGEN SATURATION (FROM 100% TO 75%) IN BLOOD PERFUSING THE BRAIN WILL CAUSE:
    A. No change in the rate of cerebral circulation
    B. Vasoconstriction of cerebral arterioles
    C. Vasodilation and an increase in the rate of cerebral circulation
    D. A significant increase in neuronal excitability
                Ref. p. 368

258. THE CEREBRAL BLOOD FLOW (IN ml blood/min/100 grams BRAIN TISSUE) IN HYPERTENSIVE PATIENTS IS:
    A. Greater than in normotensive individuals
    B. Less than in normotensive individuals
    C. Equal as in normotensive individuals
    D. Less than in persons suffering a hypotensive episode
                Ref. Fig. 32-3, p. 368

259. WHEN THE RATE OF VENOUS INFUSION OF BROMOSULFOPHTHALEIN (mg/min) IN A HUMAN EQUALS THE RATE OF EXTRACTION OF BROMOSULFOPHTHALEIN (BSP) BY THE LIVER (Ex), THE CONCENTRATION OF BSP (IN mg/ml) IN A PERIPHERAL BRACHIAL VEIN WILL NOT CHANGE; THIS WILL BE EQUAL TO THAT IN ALL ARTERIES, INCLUDING THE HEPATIC ARTERY (a). IF SIMULTANEOUSLY A HEPATIC VENOUS SAMPLE (v) CAN BE OBTAINED, THE (a)-(v) DIFFERENCE CAN BE CALCULATED. THE RATE OF BLOOD FLOW THROUGH THE LIVER UNDER THESE CIRCUMSTANCES CAN BE CALCULATED TO BE EQUAL TO:
    A. $\dfrac{(a)-(v)}{Ex}$
    B. $(a)-(v) \times Ex$
    C. $\dfrac{Ex}{(a)-(v)}$
    D. $\dfrac{Ex}{\dfrac{1}{(a)-(v)}}$
                Ref. p. 370

SECTION I - THE CARDIOVASCULAR SYSTEM

260. SYMPATHETIC STIMULATION CAUSES A:
    A. Discrete vasoconstriction restricted to the small intestine while there is concomitant vasodilation of salivary glands and gastric vessels
    B. Generalized vasodilation which is coupled with the inhibition of smooth muscle contraction
    C. Generalized constriction of vessels in the gastrointestinal tract including salivary glands and pancreas
    D. Selective vasoconstriction of the salivary gland and pancreas while there is vasodilation of the gastrointestinal tract
    Ref. p. 370

261. ASCITES IS LEAST LIKELY TO OCCUR IF OBSTRUCTION TO PORTAL BLOOD FLOW OCCURS AT THE LEVEL OF THE:
    A. Suprahepatic veins
    B. Liver sinusoids
    C. Intrahepatic periportal venules
    D. Portal vein (liver hilus)   Ref. p. 371

262. THE LARGE INCREASE IN BLOOD FLOW THAT TAKES PLACE IN SKELETAL MUSCLE THAT PERFORMS EXERCISE IS DUE TO ALL FACTORS LISTED BELOW, BUT PRIMARILY TO:
    A. Central aortic hypertension
    B. Sympathetic stimulation directly to arterioles in muscle tissue (vasodilator fibers)
    C. Local phenomena due to increased metabolic activity (diminished oxygen accumulated metabolites)
    D. Circulating epinephrine (which acts on beta receptors of vessels)
    Ref. p. 373

263. THE AREA OF THE CENTRAL NERVOUS SYSTEM DIRECTLY CONCERNED WITH THE REGULATION OF BLOOD FLOW THROUGH THE SKIN AND WHICH RESPONDS TO TEMPERATURE VARIATIONS IS LOCATED IN THE:
    A. Vestibular nuclei of the brain stem
    B. Mammillary bodies of the hypothalamus
    C. Preoptic nuclei of the hypothalamus
    D. Parieto-occipital portion of the cerebral cortex
    E. Nucleus dentatus of the cerebellum   Ref. p. 375

264. CUTANEOUS VASOCONSTRICTION IS MEDIATED BY AN INCREASED FIRING OF THE SYMPATHETIC FIBERS TO THE SKIN. CUTANEOUS VASODILATION IS MEDIATED BY:
    A. Myelinated motor fibers
    B. Parasympathetic cholinergic fibers to the skin
    C. Sympathetic fibers
    D. No nervous influence, but by a purely local hormonal mechanism
    Ref. pp. 375-376

265. A PERSON STANDING UP AFTER SITTING WITH ONE LEG CROSSED OVER THE OTHER DEVELOPS A RED PATCH OVER THE SKIN THAT WAS PREVIOUSLY COMPRESSED. THIS IS:
    A. Due to reactive hyperemia
    B. Known as a wheal
    C. Known as the red flare
    D. Due to released myoglobin   Ref. p. 376

SECTION I - THE CARDIOVASCULAR SYSTEM   55

266. VASOCONSTRICTION IN THE SKIN:
   A. Does not play any role except during exposure to cold weather when the body must conserve heat
   B. Is practically of little importance in shifts of blood from one vascular territory to another
   C. Is restricted to the epidermal plexus and does not involve the subcutaneous plexus because of the special arterio-venous anastomoses found here
   D. May shift 10% of the blood volume from the skin into the heart and large vessels   Ref. p. 376

267. A PROMINENT CLINICAL COMPONENT OF RAYNAUD'S DISEASE IS:
   A. Acute vasodilation of vessels of the hands when exposed to cold weather
   B. A surprising sparing of symptoms and signs in the feet which do not share the phenomena observed in the hands
   C. Acromegalia
   D. A severe pain in the extremities coincidental with vasoconstriction
          Ref. p. 377

268. BUERGER'S DISEASE IS CHARACTERIZED BY ALL OF THE FOLLOWING FEATURES, EXCEPT:
   A. Intermittent claudication of the lower extremities
   B. Hypersensitivity to nicotine
   C. Hypersensitivity to circulating epinephrine
   D. Occurs in both upper and lower extremities
          Ref. p. 377

SECTION II - PHYSIOLOGY OF THE BODY FLUIDS AND THE KIDNEYS

FOR EACH OF THE FOLLOWING MULTIPLE CHOICE QUESTIONS, SELECT THE ONE MOST APPROPRIATE ANSWER:

269. ALL STATEMENTS BELOW ARE TRUE, EXCEPT:
   A. In an adult lean male, water composes about 57% of the total body weight
   B. In a newborn infant the ratio of solids/water is much less than in adults
   C. With advancing age, the percentage of body water in relation to total body weight decreases
   D. Total body water increases significantly in obese persons
   E. More than half of the total body water is intracellular
   Ref. p. 380

270. A USEFUL FIGURE TO REMEMBER IS THAT OF THE BLOOD VOLUME IN RELATION TO THE TOTAL BODY WEIGHT. IN LEAN PERSONS IT EQUALS APPROXIMATELY:
   A. 40 ml blood/kg body weight
   B. 80 ml blood/kg body weight
   C. 100 ml blood/kg body weight
   D. 120 ml blood/kg body weight
   E. 150 ml blood/kg body weight
   Ref. p. 381

271. IN CALCULATIONS CONCERNED WITH REPLACEMENT OF BLOOD IN AN OBESE FEMALE ONE SHOULD REMEMBER THAT:
   A. Adipose tissue is a highly hydrated tissue
   B. An obese female has a lower blood volume per total body weight than lean persons
   C. An obese female has a higher blood volume per total body weight than lean persons
   D. Obese females always have a low hematocrit
   Ref. p. 381

272. PERSONS WHOSE HEMATOCRIT EQUALS 60 HAVE, WITH RESPECT TO NORMALS, A:
   A. Supersaturated hemoglobin
   B. Higher oxygen tension in their blood
   C. Greater oxygen consumption
   D. Lowered peripheral resistance
   E. Higher viscosity of blood
   Ref. p. 382

273. THE VALUE OF HEMATOCRIT MEASURED IN BLOOD SAMPLED FROM A VEIN IN THE FLEXURE OF THE ARM:
   A. Equals the capillary hematocrit
   B. Has a higher value than capillary hematocrit
   C. Has a lower value than whole body hematocrit
   D. Owes its high value to the fact that the $\frac{\text{surface area}}{\text{volume}}$ ratio is greater in the case of veins as compared to that found in capillaries
   Ref. p. 382

SECTION II - PHYSIOLOGY OF THE BODY FLUIDS AND THE KIDNEYS

274. THE REASON WHY THE T-1824 (EVANS BLUE) SPACE SEEMS TO INCREASE OVER A PERIOD OF TWO HOURS, LIES IN THE FACT THAT THE BLUE DYE:
    A. Enters the red blood cells and disappears from plasma
    B. Passes the renal glomeruli and is not reabsorbed by the tubules at a rapid rate
    C. Is rapidly metabolized by the liver
    D. Is attached to plasma proteins which leave the circulation at a slow but steady rate through capillary pores
    E. Has a varying spectrophotometric absorption peak with time
    Ref. p. 383

275. ALL OF THE FOLLOWING SUBSTANCES CAN BE USED TO MEASURE THE EXTRACELLULAR FLUID SPACE, WITH THE EXCEPTION OF:
    A. Radioactive serum albumin (RISA)
    B. Radioactive sodium
    C. Thiosulfate ion
    D. Inulin
    E. Thiocyanate ion
    Ref. p. 384

276. THE DEGREE OF INTRACELLULAR PENETRATION OF MOLECULES OF EACH OF THE FOLLOWING SUBSTANCES VARIES. THE VOLUME OF DISTRIBUTION IS GREATER FOR:
    A. Albumin
    B. Potassium
    C. Sodium
    D. Chloride
    E. Inulin
    Ref. p. 384

277. THE FOLLOWING MEASUREMENTS WERE MADE IN AN ADULT HUMAN:
    Total body water: 40 liters of water
    Plasma water: 2.5 liters
    Red blood cells mass: 1.5 liters
    Extracellular fluid water: 15 liters
    ONE MAY CONCLUDE FROM THESE FIGURES THAT THE "INTERSTITIAL" FLUID VOLUME IS APPROXIMATELY EQUAL TO:
    A. 11.5 liters
    B. 12.5 liters
    C. 14.5 liters
    D. 15.5 liters
    E. 16.5 liters
    Ref. p. 385

278. AMONG NON-ELECTROLYTE SUBSTANCES THE MAJOR CONSTITUENT (IN MILLIGRAMS PER 100 ml) IN THE EXTRACELLULAR FLUID IS REPRESENTED BY MOLECULES OF:
    A. Phospholipids
    B. Cholesterol
    C. Neutral fat
    D. Glucose
    E. Urea
    Ref. Fig. 33-6, p. 385

279. THE MAJOR OSMOLAR CONTRIBUTION IN PLASMA IS GIVEN BY IONIZED ATOMS OF:
    A. Potassium
    B. Chloride
    C. Phosphate
    D. Magnesium
    E. Sodium
    Ref. Fig. 33-5, p. 385

SECTION II - PHYSIOLOGY OF THE BODY FLUIDS AND THE KIDNEYS

280. QUANTITATIVELY, THE MAJOR ANION IN THE INTRACELLULAR FLUID IS THAT OF:
    A. Protein
    B. Sulfate
    C. Phosphate
    D. Chloride
    E. Bicarbonate
    Ref. Fig. 33-5, p. 385

281. IN THE PHENOMENON OF OSMOSIS:
    A. Water fluxes are predominant from the space where solutes are less concentrated to that where they are more concentrated
    B. Water fluxes are predominant from the space where solutes are more concentrated to that where they are less concentrated
    C. Small ions of either sodium or chloride have a greater "activity" than a similar quantity of large ions of protein
    D. Cellular membranes oppose a definite resistance to the establishment of equilibrium of water fluxes as opposed to more permeable capillary membranes
    Ref. p. 386

282. OSMOTIC FORCES ARE ESSENTIALLY DUE TO A(N):
    A. Excess of chemical activity of solutes where solute molecules are in excess, thus attracting water with greater force
    B. Excess of chemical activity of molecules of water where the solution is more dilute
    C. Greater permeability for water on one side of the membrane due to the structural composition of the membrane
    D. Higher atmospheric pressure where the solution is more dilute
    Ref. p. 386

283. ALL STATEMENTS LISTED BELOW ARE TRUE, EXCEPT:
    A. Calcium ions exert twice the osmotic force of sodium ions
    B. A molecule of albumin with a molecular weight equal to 70,000 exerts an osmotic effect identical to a sodium ion which has an atomic weight equal to 23
    C. Half the gram molecular weight of sodium chloride will provide the activity of one osmol
    D. The osmotic pressure in mm Hg equals the concentration in milliosmols per liter times 19.3
    E. The ratio intracellular osmolality cannot vary from 1.0 in the living
         extracellular osmolality
    individual for more than a few seconds
    Ref. p. 387

284. THE RATIO   OSMOTIC PRESSURE PLASMA   EQUALS:
                OSMOTIC PRESSURE INTERSTITIAL FLUID
    A. 1.0 at all times
    B. Is slightly greater than 1.0 at all times
    C. Is slightly less than 1.0 at all times
    D. May fluctuate with periods of fasting and feeding
    Ref. p. 387

SECTION II - PHYSIOLOGY OF THE BODY FLUIDS   59
                  AND THE KIDNEYS

285. THE REASON FOR THE CORRECT STATEMENT IN THE
     PREVIOUS QUESTION LIES IN THE FACT THAT:
     A. There is the same concentration of osmotic particles in plasma and
        interstitial fluid
     B. There is an excess of protein in plasma which does not leave the
        vascular compartment
     C. The Donnan effect compensates for the excess of protein
     D. Absorption of glucose and amino acids into the plasma space is not
        associated with water fluxes from and into cells rapidly enough
                                          Ref. p. 387

286. IF ONE CONSIDERS THAT THE TOTAL OSMOLAR CONCENTRATION IN
     RED BLOOD CELLS EQUALS APPROXIMATELY 300 MILLIOSMOLS/LITER
     AND THAT THIS REPRESENTS (300 x 19.3 = 5790 mm Hg) 5,790 mm Hg,
     ONE MAY CONCLUDE THAT THE FORCE WHICH RUPTURES THE MEM-
     BRANE OF A RED CELL PLACED IN DISTILLED WATER EQUALS:
     A. 5.8 atmospheres   D. 8.8 atmospheres
     B. 6.8 atmospheres   E. 9.8 atmospheres
     C. 7.8 atmospheres   Ref. p. 388

287. A MAN IN THE DESERT DRINKS QUICKLY 5 LITERS OF DISTILLED
     WATER TO RELIEVE HIS THIRST. WE ASSUME THAT THIS VOLUME
     RAPIDLY WAS ABSORBED FROM THE GASTROINTESTINAL TRACT,
     AND THAT NO URINE WAS FORMED DURING THE FIRST 5 MINUTES
     AFTER DRINKING THE WATER. BEFORE RELIEVING THIS THIRST HIS
     BODY COMPOSITION WAS AS FOLLOWS:
            TOTAL BODY WATER   35 liters
            EXTRACELLULAR WATER   12 liters
            INTRACELLULAR WATER   23 liters
            PLASMA OSMOLARITY   300 mOsmols/liter
     AS SOON AS THE WATER WAS ABSORBED FROM THE GASTROINTES-
     TINAL TRACT, AND IMMEDIATELY AFTER OSMOTIC EQUILIBRIUM
     OCCURRED BETWEEN INTRA AND EXTRACELLULAR SPACES, THE
     OSMOTIC CONCENTRATION IN PLASMA BECAME EQUAL TO:
     A. 222 mOsm/Liter
     B. 242 mOsm/Liter
     C. 262 mOsm/Liter
     D. 282 mOsm/Liter
     E. 292 mOsm/Liter   Ref. p. 389

288. OF THE FIVE LITERS OF WATER INGESTED BY THE SUBJECT OF THE
     PREVIOUS QUESTION, THE VOLUME OF EXTRA WATER WHICH EN-
     TERED THE CELLS AT EQUILIBRIUM EQUALED:
     A. 1.25 liters
     B. 2.75 liters
     C. 3.25 liters
     D. 4.00 liters
     E. 5.25 liters   Ref. p. 389

SECTION II - PHYSIOLOGY OF THE BODY FLUIDS AND THE KIDNEYS

289. BY MISTAKE A ONE LITER BOTTLE CONTAINING HYPERTONIC SALINE (1,500 mOsm NaCl per liter) WAS ADMINISTERED INTRAVENOUSLY TO A PATIENT RAPIDLY, INSTEAD OF USING ISOTONIC SALINE. THE BODY COMPOSITION OF THE PATIENT BEFORE THE INFUSION WAS:
    TOTAL BODY WATER   35 liters
    EXTRACELLULAR WATER   12 liters
    INTRACELLULAR WATER   23 liters
    PLASMA OSMOLARITY   300 mOsmoles/liter
    IF NO INTRACELLULAR WATER HAD DIFFUSED OUT OF THE CELLS THE PLASMA OSMOLARITY WOULD HAVE RISEN TO VALUES EQUAL TO:
    A. 342 mOsm/Liter
    B. 352 mOsm/Liter
    C. 372 mOsm/Liter
    D. 392 mOsm/Liter
    E. 402 mOsm/Liter
    Ref. p. 390

290. BECAUSE THE INTRACELLULAR WATER DIFFUSED OUT OSMOTICALLY, THE NEW PLASMA OSMOLARITY, AT EQUILIBRIUM, WAS EQUAL TO:
    A. 324 mOsm/Liter
    B. 334 mOsm/Liter
    C. 342 mOsm/Liter
    D. 348 mOsm/Liter
    E. 350 mOsm/Liter
    Ref. p. 390

291. THE VOLUME OF WATER THAT DIFFUSED OUT OF THE CELLULAR COMPARTMENT TO DILUTE THE EXTRACELLULAR FLUID, AND THUS PREVENT THE OSMOLARITY OF PLASMA FROM RISING TO LETHAL LEVELS WAS EQUAL TO:
    A. 0.27 liters
    B. 1.27 liters
    C. 3.27 liters
    D. 4.27 liters
    E. 5.27 liters
    Ref. p. 390

292. BY INJECTING HYPERTONIC GLUCOSE OR UREA INTRAVENOUSLY IN A PATIENT WHO HAS CEREBRAL EDEMA ONE MAY CAUSE A SUBSTANTIAL DECREASE OF INTRACELLULAR WATER WITHIN:
    A. Five seconds
    B. Five minutes
    C. Fifty minutes
    D. Five hours
    Ref. Fig. 33-8, p. 390

293. IN A PATIENT SIMILAR TO THAT DESCRIBED IN THE PREVIOUS QUESTION, THE LOSS OF BODY WATER THROUGH DIURESIS FOLLOWS THE INFUSION OF HYPERTONIC MANNITOL:
    A. Five hours after the intracellular water diffused out to the extracellular fluid space
    B. And can amount to several liters of excess body water
    C. And should be followed by infusion of an equal volume of dextrose and water
    D. Will nevertheless leave a permanent expansion of the extracellular fluid space as mannitol is avidly retained in the plasma space
    Ref. p. 391

SECTION II - PHYSIOLOGY OF THE BODY FLUIDS AND THE KIDNEYS

294. IN A NORMAL PERSON, WATER INJECTED (2 LITERS) TOGETHER WITH GLUCOSE IN ISOTONIC CONCENTRATION WILL:
    A. Be retained predominantly in the extracellular fluid space
    B. Be removed by the kidney within 2 to 3 hours
    C. Be retained in the intracellular fluid space
    D. Expand both the extra- and intracellular fluid spaces for at least 24 hours   Ref. p. 391

295. AN IMPORTANT DIFFERENCE BETWEEN ISOTONIC SALINE AND RINGER'S SOLUTION IS THAT THE LATTER:
    A. Contains magnesium ions
    B. Is hypertonic with respect to the first one
    C. Has a higher concentration of sodium ions
    D. Contains potassium and calcium ions
    E. Contains albumin   Ref. p. 391

296. THE LONGEST THIN LOOPS OF HENLE ARE PREDOMINANTLY A PART OF NEPHRONS WHOSE GLOMERULI ARE LOCATED IN THE:
    A. Cortex
    B. Papilla
    C. Juxtamedullary region of the cortex
    D. Medulla   Ref. p. 394

297. THE VOLUME OF BLOOD FLOWING THROUGH BOTH KIDNEYS EQUALS APPROXIMATELY (IN A 70 kg MAN, LYING QUIETLY IN BED, WITHOUT ANY ANXIETY):
    A. 0.2 liters/minute
    B. 0.5 liters/minute
    C. 0.8 liters/minute
    D. 1.2 liters/minute
    E. 2.00 liters/minute
    Ref. p. 394

298. THE DROP IN PRESSURE FROM THE GLOMERULAR CAPILLARIES TO THE PERITUBULAR CAPILLARIES KNOWN AS VASA RECTA AMOUNTS TO APPROXIMATELY:
    A. 5 mms Hg
    B. 10 mms Hg
    C. 20 mms Hg
    D. 40 mms Hg
    E. 60 mms Hg
    Ref. Fig. 34-4, p. 395

299. THE GLOMERULAR CAPILLARY HAS ITS ENDOTHELIAL LAYER LYING ON A BASEMENT MEMBRANE, WHICH IN TURN IS SUPPORTED BY AN EPITHELIAL LAYER. UNDER NORMAL CONDITIONS, ITS PERMEABILITY:
    A. Allows most protein molecules to pass through without severe hindrance
    B. Does not allow red cells or leucocytes to pass through
    C. Indirectly (Donnan effect) determines a lower concentration of chloride and bicarbonate ions in the ultrafiltrate of the Bowman's space
    D. Indirectly determines a lower concentration of glucose and creatinine in the ultrafiltrate of the Bowman's space
    Ref. p. 396

SECTION II - PHYSIOLOGY OF THE BODY FLUIDS AND THE KIDNEYS

IN THE FOLLOWING QUESTIONS (300-319) A STATEMENT IS FOLLOWED BY FOUR POSSIBLE ANSWERS. ANSWER BY USING THE KEY OUTLINED BELOW:
A. If only 1, 2 and 3 are correct
B. If only 1 and 3 are correct
C. If only 2 and 4 are correct
D. If only 4 is correct
E. If all are correct

300. IF THE FOLLOWING VALUES ARE OBTAINED BY PHYSIOLOGICAL PROCEDURES IN A 70 KILOGRAM MAN:
    Renal blood flow   1.2 liters
    Renal plasma flow   0.6 liters
    Glomerular filtration rate 0.120 liters
    Hematocrit   0.50
    Extraction ratio of PAH   0.90
    Tm glucose   336 mg/minute
    1. The filtration fraction equals 0.20
    2. There is probably glucosuria when the plasma glucose equals 0.9 mg/ml
    3. The glomerular filtration rate is normal
    4. Arterial blood is totally exposed to renal tubules
    Ref. p. 396

301. UNDER NORMAL CONDITIONS:
    1. The colloid osmotic pressure of plasma in the efferent arteriole increases by 20%
    2. The pressure in the Bowman capsule equals approximately 15 mm Hg
    3. The net filtration pressure equals approximately 25 mm Hg
    4. An increase in the afferent arteriolar tone decreases the net glomerular filtration pressure   Ref. p. 397

302. AN INCREASED FIRING RATE OF RENAL SYMPATHETIC FIBERS:
    1. Will decrease renal blood flow by not more than 5% of the control values
    2. Causes little change in filtration rate when the stimulus is mild
    3. Does not occur when humans are passively tilted from the horizontal to the vertical position
    4. Can give rise to hypertension   Ref. p. 398

303. AN INCREASE IN MEAN AORTIC BLOOD PRESSURE FROM 100 TO 200 mm Hg WILL CAUSE:
    1. Afferent arteriolar dilation with diminished resistance
    2. A significant increase in renal plasma flow
    3. A significant increase in filtration fraction
    4. A slight (5%) increase in the glomerular filtration rate
    Ref. p. 398

304. ACTIVE TRANSPORT OF SODIUM IN THE KIDNEY:
    1. Occurs only on the luminal side of the proximal tubular cells and not on the basal side (capillary side)
    2. Decreases the concentration of sodium inside the cell
    3. Acts to increase the potential of the tubular lumen by 20 mV
    4. Creates an electrical potential of -70 mV inside the proximal tubular cell   Ref. p. 399

SECTION II - PHYSIOLOGY OF THE BODY FLUIDS AND THE KIDNEYS

63

305. THE ELECTRICAL DIFFERENCE BETWEEN TUBULAR LUMEN AND INTRACELLULAR SPACE IN THE PROXIMAL TUBULAR CELLS OF THE KIDNEY:
   1. Equals about 70 mV
   2. Opposes the chemical potential determined by the concentration gradient of sodium between the two surfaces of the apical membrane of cells
   3. Is sufficient to stop the active transport of sodium under certain circumstances
   4. Favors the flux of sodium through the brush border of the apical surface Ref. p. 399

306. IN PROXIMAL TUBULAR CELLS, THERE IS IN ADDITION TO THE MECHANISM WHICH TRANSPORTS SODIUM ACTIVELY, ANOTHER INDEPENDENT ACTIVE TRANSPORT SYSTEM FOR THE REABSORPTION OF:
   1. Glucose
   2. Calcium ions
   3. Amino acids
   4. Potassium ions Ref. p. 399

307. THE PERMEABILITY OF THE TUBULAR CELL MEMBRANE IN THE PROXIMAL TUBULE:
   1. Is much less for water than for urea
   2. Allows water to diffuse rapidly after the actively reabsorbed sodium
   3. Is high for polymers of fructose, such as inulin
   4. Is zero for sucrose Ref. p. 400

308. THE ELECTRICAL DIFFERENCE BETWEEN TUBULAR LUMEN AND PERITUBULAR FLUID:
   1. Is equal to approximately 20 mV in the proximal tubule
   2. Is equal to 60 mV in the distal tubule
   3. Increases to twice the control value in the distal tubule when sodium reabsorption occurs rapidly
   4. Offers an electrical impedance to the diffusion of chloride, phosphate and bicarbonate Ref. p. 400

309. POTASSIUM IONS:
   1. Decrease in concentration, progressively, in the proximal tubule's lumen
   2. Are secreted by the distal tubule
   3. "Exchange" for sodium ions in the distal tubule under the influence of aldosterone
   4. Passively diffuse through distal tubular cells from plasma to lumen aided by the electrical potential difference
   Ref. pp. 400, 403, 419

310. THE DISTAL TUBULAR CELLS:
   1. Are less permeable for most substances than the proximal tubular cells
   2. Develop a much smaller electrochemical potential difference than proximal cells
   3. Reabsorb 12-15% of the filtrate under normal basal conditions
   4. Allow a more rapid and easy diffusion of sodium ions than the proximal tubular cells Ref. p. 401

SECTION II - PHYSIOLOGY OF THE BODY FLUIDS AND THE KIDNEYS

311. THE PASSAGE OF WATER FROM TUBULAR LUMEN TO THE PERITUBULAR CAPILLARIES:
 1. Occurs by active transport in the proximal tubule
 2. Occurs from a hypertonic lumen to a hypotonic plasma in osmotic diuresis
 3. Creates an electrical potential, negative in plasma, positive in the lumen
 4. Occurs by osmotic diffusion (solutes transported first, leaving behind a hypotonic solution, creating an osmotic gradient) which "pulls" water after the solutes   Ref. p. 401

312. REABSORPTION OF PROTEIN BY RENAL NEPHRONS:
 1. Occurs in the proximal tubule
 2. Takes place by pinocytosis
 3. Leaves the tubular fluid protein-free at the end of the proximal tubule
 4. May recover up to 30 grams of protein filtered per day
   Ref. p. 402

313. NO REABSORPTION AT ALL TAKES PLACE FROM THE TUBULAR FLUID AND INTO THE PERITUBULAR CAPILLARIES IN THE CASE OF:
 1. Creatinine
 2. Para-amino-hippuric acid
 3. Inulin
 4. Phosphate   Ref. p. 402

314. THE RATIO $\frac{\text{CONCENTRATION IN URINE}}{\text{CONCENTRATION IN PLASMA}}$ IS GREATER THAN 1.0 FOR THE FOLLOWING SUBSTANCES:
 1. Creatinine
 2. Amino acids
 3. Urea
 4. Bicarbonate   Ref. Fig. 34-11, p. 403

315. BICARBONATE IONS:
 1. Are reabsorbed with difficulty by tubular cells
 2. Form carbon dioxide and water when hydrogen ions are secreted by the tubular cells into the lumen
 3. Will appear in the urine if there is no hydrogen ion secretion by the tubular cells
 4. In the Bowman's capsule have a concentration which is 5% higher than within the glomerular capillary lumen
   Ref. pp. 396, 403

316. IN THE MEDULLA OF THE KIDNEYS:
 1. Sodium is actively reabsorbed out of the lumen of the ascending portion and into the inter-tubular space
 2. The blood flow through the vasa recta is very fast due to its straight configuration
 3. Sodium removed by active transport from the ascending part of the loop diffuses back into the lumen of the descending part of the loop
 4. The concentration gradient of sodium between the initial part of the loop and the bend is much less than the gradient established between ascending and descending portions placed at the same level with respect to the base of the pyramids   Ref. pp. 407-408

SECTION II - PHYSIOLOGY OF THE BODY FLUIDS AND THE KIDNEYS

317. VASA RECTA:
 1. Collect all the water that diffuses out of the ascending part of the loop of Henle
 2. Have an intraluminal pressure equal to 50 mm Hg
 3. Have a thick muscular coat capable of vigorous vasomotion
 4. Have an extremely poor efficiency to carry sodium away from the medulla Ref. Fig. 34-4, p. 408

318. AS THE TUBULAR FLUID ENTERS THE DISTAL TUBULE AFTER LEAVING THE ASCENDING PORTION OF THE LOOP OF HENLE:
 1. It is hypoosmotic with respect to plasma
 2. It is subject to changing permeability of the tubular cells in accordance to the levels of antidiuretic hormone
 3. It contains creatinine at concentrations equal to about 7 times greater than in the Bowman's capsule
 4. It does not lose or gain any potassium in the remainder of the course through the renal cortex Ref. pp. 403, 408

319. IN THE COLLECTING DUCTS:
 1. The tubular fluid may increase its osmolality to a maximal value of 600 milliosmols per liter
 2. Antidiuretic hormone has no specific effects comparable to those found in the distal tubule
 3. Permeability to water under the best circumstances is relatively slow
 4. The ratio $\frac{\text{urine osmolality}}{\text{plasma osmolality}}$ may reach a value equal to 4.0
  Ref. p. 408

SECTION II - PHYSIOLOGY OF THE BODY FLUIDS AND THE KIDNEYS

SELECT THE ONE MOST APPROPRIATE ANSWER:

PLOTS OF URINARY/PLASMA RATIOS OF CONCENTRATION FOR FOUR SUBSTANCES ARE SHOWN BELOW, WHERE THESE RATIOS CHANGE AS A FUNCTION OF THE LEVEL OF RENAL NEPHRONS: THE PROXIMAL TUBULE, THE LOOP OF HENLE, THE DISTAL TUBULE AND THE COLLECTING DUCT:

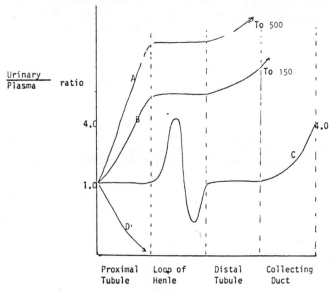

Redrawn with modifications from Figure 34-11 and Fig.35-2 in Guyton's Textbook of Medical Physiology, 4th ed. W.B. Saunders Co.

320. THE PLOT THAT BEST DESCRIBES THE CHANGES OCCURRING WITH INJECTED PARA-AMINO-HIPPURATE (AT LOW PLASMA CONCENTRATIONS) IS THE PLOT LABELED:
   A. ___
   B. ___
   C. ___
   D. ___
   Ref. Fig. 34-11, p. 403 and Fig. 35-2, p. 409

321. THE PLOT THAT BEST DESCRIBES THE CHANGES OCCURRING WITH ENDOGENOUS CREATININE IS THE PLOT LABELED:
   A. ___
   B. ___
   C. ___
   D. ___
   Ref. Fig. 34-11, p. 403

SECTION II - PHYSIOLOGY OF THE BODY FLUIDS AND THE KIDNEYS

322. THE PLOT THAT BEST DESCRIBES CHANGES OF OSMOTIC PARTICLES IN A DEHYDRATED HUMAN IS THE PLOT LABELED:
    A. ___
    B. ___
    C. ___
    D. ___   Ref. Fig. 35-2, p. 403

323. THE CURVE THAT BEST DESCRIBES CHANGES OCCURRING WITH GLUCOSE MOLECULES IN A NON-DIABETIC INDIVIDUAL IS LABELED:
    A. ___
    B. ___
    C. ___
    D. ___   Ref. Fig. 34-11, p. 403

324. OBLIGATORY URINE OUTPUT:
    A. Is due to the presence of urea and creatinine in the glomerular filtrate
    B. Is equal to at least 800 ml per day
    C. Will not take place if there is a severe dehydration
    D. Decreases with an increase in afferent arteriolar pressure
                                Ref. p. 410

325. IF IN A HUMAN THAT RECEIVES AN INFUSION OF INULIN AT A STEADY RATE, ANALYSIS REVEALS THAT PLASMA INULIN EQUALS 0.2 mg/ml, URINARY INULIN CONCENTRATION EQUALS 12 mg/ml AND THAT THERE IS A RATE OF URINE FLOW EQUAL TO 2.0 ml PER MINUTE, THE GLOMERULAR FILTRATION RATE EQUALS:
    A. 50 ml/min
    B. 120 ml/min
    C. 180 ml/min
    D. 200 ml/min   Ref. p. 411

THE FOLLOWING GRAPH DESCRIBES THE VARIATIONS OF TWO VARIABLES AS A FUNCTION OF INCREASES IN ARTERIAL BLOOD PRESSURE. IF THE POSSIBLE VARIABLES WERE:
    A. Extraction ratio of inulin
    B. Filtration fraction
    C. Glomerular filtration
    D. Renal plasma flow

Redrawn with modifications from Figure 35-3 in Guyton's Textbook of Medical Physiology, 4th ed. W.B.Saunders Co.

326. WHICH OF THE ABOVE MENTIONED VARIABLES IS DESCRIBED BY THE CURVE MARKED WITH THE NUMBER 1?:
    A. ___
    B. ___
    C. ___
    D. ___   Ref. Fig. 35-3, p. 409

SECTION II - PHYSIOLOGY OF THE BODY FLUIDS
AND THE KIDNEYS

327. WHICH OF THE ABOVE MENTIONED VARIABLES IS DESCRIBED BY THE CURVE MARKED WITH THE NUMBER 2 ?:
A. ___
B. ___
C. ___
D. ___
Ref. Fig. 35-3, p. 409

328. THE CLEARANCE OF PARA-AMINO-HIPPURATE (PAH), AT LOW CONCENTRATIONS OF PAH IN PLASMA:
A. Amounts to about 90% of the total plasma that flows through the kidneys
B. Is due solely to tubular transport
C. Can never become totally efficient due to strong affinities of erythrocytes for PAH molecules
D. Equals about 10 times the clearance of inulin
Ref. p. 412

329. IF THE FOLLOWING VALUES ARE OBTAINED IN A MAN:
Plasma PAH concentration = 0.0175 mg/ml
Clearance inulin = 100 ml/min
Urine flow rate = 2.0 ml/min
Urinary PAH concentration = 4 mg/ml
THE FILTERED LOAD OF PAH EQUALS:
A. 0.0175 mg/minute
B. 0.175 mg/minute
C. 1.75 mg/minute
D. 17.5 mg/minute
E. 175 mg/minute
Ref. p. 412

330. IF THE VALUES QUOTED IN THE PREVIOUS QUESTION ARE USED TO CALCULATE THE CLEARANCE OF PAH, IT WOULD BE CORRECT TO SAY THAT IT IS EQUAL TO:
A. 4.57 ml/min
B. 45.7 ml/min
C. 457 ml/min
D. 4570 ml/min
Ref. p. 411

331. THE MAXIMAL CAPACITY TO REABSORB GLUCOSE WILL BE EXCEEDED IF THE:
A. Tubular load for glucose/Tm glucose ratio exceeds 1.0
B. Tubular load for glucose/Tm glucose ratio is under 1.0
C. Tm for glucose equals 320 mg/minute
D. Glomerular filtration rate is doubled and the plasma glucose is diminished to one-half the normal values
Ref. p. 412

332. IN THE NORMAL ADULT HUMAN, THE PLASMA GLUCOSE EQUALS 1.0 mg/ml DURING FASTING PERIODS, THE GLOMERULAR FILTRATION RATE EQUALS ABOUT 120 ml/minute AND THE Tm FOR GLUCOSE EQUALS 320 mg/minute. THIS MEANS THERE IS A RESERVE OF TUBULAR REABSORPTION OF GLUCOSE EQUAL TO:
A. 0.67 times the basal filtered load
B. 1.67 times the basal filtered load
C. 2.67 times the basal filtered load
D. 3.67 times the basal filtered load
E. 4.67 times the basal filtered load
Ref. Fig. 35-5, p. 413

SECTION II - PHYSIOLOGY OF THE BODY FLUIDS AND THE KIDNEYS

333. OF ALL THE FOLLOWING SUBSTANCES, ONE HAS NO TRANSPORT MAXIMUM AS TUBULES AVIDLY REABSORB EVER INCREASING FILTERED LOADS. SUCH A SUBSTANCE IS:
    A. Glucose
    B. Urate
    C. Lactate
    D. Sodium
    E. Phosphate
    Ref. p. 413

334. OF THE FOLLOWING MEASUREMENTS, ONE IS INDISPENSABLE TO CALCULATE Tm (TRANSPORT MAXIMUM) FOR ALL SUBSTANCES. SELECT THE INDISPENSABLE ONE:
    A. Clearance of inulin
    B. Extraction ratio of inulin
    C. Clearance of PAH
    D. Extraction ratio of PAH
    E. Total renal plasma flow
    Ref. p. 413

335. THE FUNDAMENTAL REASON WHY AN EXTREMELY LOW POTASSIUM CONCENTRATION IN PLASMA MAY BE DANGEROUS IN A HUMAN IS BECAUSE SUCH DEGREE OF HYPOKALEMIA:
    A. Increases the resting membrane potential of myocardial fibers
    B. Decreases the resting membrane potential of myocardial sinoatrial cells
    C. Is likely to be associated with tetanic contractions (laryngeal spasm)
    D. Depresses cellular permeability   Ref. p. 415

336. IN THE ABSENCE OF ALDOSTERONE:
    A. No sodium is reabsorbed by the proximal tubules
    B. The filtered load of sodium is increased
    C. The distal tubules decrease the reabsorption of the small fraction of filtered sodium that reaches them
    D. The glomerular filtration rate increases
    Ref. p. 416

337. IF A PERSON SURGICALLY DEPRIVED OF HIS ADRENALS AND NOT MEDICATED WITH STEROIDS WERE TO INGEST SUPPLEMENTARY SODIUM CHLORIDE TO BALANCE THE EXCESSIVE LOSS OF THIS SALT, HE SHOULD TAKE AT LEAST:
    A. 1 gram per day
    B. 5 grams per day
    C. 15 grams per day
    D. 50 grams per day   Ref. p. 416

338. ONE WOULD EXPECT AN INCREASED RATE OF SECRETION OF ALDOSTERONE IF A PERSON HAS, OVER A PERIOD OF SEVERAL HOURS OR DAYS:
    A. 125 mEq Na+/liter of plasma
    B. 2.5 mEq K+/liter of plasma
    C. A blood volume equal to 150 ml/Kg lean body weight
    D. A therapeutic administration of cortisol in large doses
    Ref. pp. 416-417

339. A RATHER SPECIFIC INDUCER OF THE SECRETION OF ALDOSTERONE IS:
    A. Vasopressin
    B. Cortisol
    C. Insulin
    D. Angiotensin
    E. Renin
    Ref. p. 417

SECTION II - PHYSIOLOGY OF THE BODY FLUIDS AND THE KIDNEYS

340. IF HYPERKALEMIA IS INDUCED BY INFUSION OF LARGE AMOUNTS OF POTASSIUM CHLORIDE DURING A SHORT PERIOD OF TIME, BUT THE KIDNEYS AND THE ADRENAL CORTICES ARE NORMAL:
    A. The rate of sodium excretion will decrease (but none of the other effects)
    B. The extracellular fluid space will increase (but none of the other effects)
    C. The potassium excretion will increase (but none of the other effects)
    D. The rate of aldosterone secretion will increase (but none of the other effects)
    E. All of the above   Ref. p. 418

341. AN ACID pH OF THE EXTRACELLULAR FLUID CAUSES A(N):
    A. Increased renal reabsorption of the filtered bicarbonate
    B. Increased renal reabsorption of the filtered chloride
    C. Decreased renal reabsorption of bicarbonate
    D. Excessive loss of potassium ions from the proximal tubule
                                  Ref. p. 419

342. THE CONCENTRATION OF PLASMA PHOSPHATE UNDER NORMAL CONDITIONS EQUALS 1.0 mM/L. ANYTIME THE PLASMA CONCENTRATION DROPS BELOW 0.8 mM/L THE KIDNEYS CEASE TO EXCRETE IT, AS THERE SEEMS TO BE A Tm FOR PHOSPHATE. IF THE RATE OF GLOMERULAR FILTRATION RATE EQUALS 0.125 LITERS PER MINUTE THE VALUE OF THIS Tm EQUALS:
    A. 0.05 mM/minute   D. 0.20 mM/minute
    B. 0.10 mM/minute   E. 0.25 mM/minute
    C. 0.15 mM/minute   Ref. p. 419

343. IN THE KIDNEY, PARATHYROID HORMONE INFLUENCES PHOSPHATE METABOLISM AS IT:
    A. Favors the tubular reabsorption of filtered phosphate
    B. Inhibits the tubular reabsorption of phosphate
    C. Decreases the tubular load by decreasing significantly the glomerular filtration rate
    D. Increases the tubular load by increasing twofold the effective renal plasma flow   Ref. p. 419

344. THE VOLUME OF WATER PRODUCED EACH DAY THROUGH METABOLISM OF CARBOHYDRATES, PROTEIN AND FAT EQUALS ABOUT:
    A. Zero
    B. 5-10 ml
    C. 30-50 ml
    D. 150-250 ml
    E. 500-1000 ml   Ref. p. 420

345. THE VOLUME OF WATER INSENSIBLY LOST (ALVEOLAR SURFACE AND DIFFUSION THROUGH THE SKIN) BY A NORMAL ADULT HUMAN EXPOSED TO AN AIR CONDITIONED ENVIRONMENT EQUALS ABOUT:
    A. 50 ml/day
    B. Half the volume excreted by the kidney (urine volume 1500 ml)
    C. 2000 ml
    D. Zero   Ref. p. 420

SECTION II - PHYSIOLOGY OF THE BODY FLUIDS AND THE KIDNEYS

346. A PERSON CONGENITALLY DEPRIVED OF SWEAT GLANDS WILL:
    A. Have no loss of heat through the skin
    B. Be "panting" all the time
    C. Lose up to 400 ml of water by diffusion through the skin
    D. Lose all insensible water through the lungs only
    Ref. p. 420

347. A PERSON ENGAGED IN A VERY ACTIVE MUSCULAR EXERCISE:
    A. May lose up to 3-4 liters of water through the skin
    B. Will have no increase in water lost through the lungs
    C. Will lose more water through the kidney because less ADH is secreted
    D. Has part of his extra heat usefully converted into work
    Ref. p. 420

348. WHEN THE RATE OF ANTIDIURETIC HORMONE SECRETION IS MAXIMAL THE RATE OF URINE FORMATION DROPS TO THE EQUIVALENT OF:
    A. Zero urine/day
    B. 10-20 ml urine/day
    C. 400-500 ml urine/day
    D. 1000-1200 ml urine/day
    Ref. p. 421

349. IT IS MOST LIKELY THAT IN RESPONSE TO A PLASMA OSMOLALITY EQUAL TO 270 mOsm/L THE NEURONS RESPONSIBLE FOR SECRETION OF ANTIDIURETIC HORMONE:
    A. Have lost part of their cellular water
    B. Are forming antidiuretic hormone at a high rate
    C. Have gained water
    D. Are not modified from their basal activity
    Ref. p. 421

350. THE MAGNITUDE OF THE DEVIATION FROM THE NORMAL PLASMA OSMOLALITY (285 mOsm/L) REQUIRED TO ACTIVATE OR DEACTIVATE THE OSMORECEPTORS AMOUNTS TO:
    A. 1-2% of the normal
    B. 5-10% of the normal
    C. 15-20% of the normal
    D. 20-30% of the normal
    Ref. p. 421

351. THE TEMPORAL SEQUENCE OF ANTIDIURETIC HORMONE (ADH) DE-ACTIVATION AND METABOLISM CAUSES THE INITIATION OF A WATER DIURESIS TO BE:
    A. Delayed 2 hours after drinking one liter of beer
    B. Delayed 30 minutes after drinking one liter of water
    C. Immediate (one or two minutes) after ingestion of 1 liter of water
    D. Far more rapid than the initiation of an osmotic diuresis
    Ref. p. 421

352. A PERSON SUFFERING DIABETES INSIPIDUS:
    A. Is curiously free of thirst
    B. Is likely to be dehydrated during prolonged sleep
    C. May have a maximal plasma osmolality equal to 285 mOsm/L
    D. Will not develop a urine osmolality of 1200 mOsm/L even if injected with ADH
    Ref. p. 422

SECTION II - PHYSIOLOGY OF THE BODY FLUIDS AND THE KIDNEYS

353. AN ANIMAL KEPT ON A SALT-FREE DIET FOR A PROLONGED TIME WILL DEVELOP A HYPOTONIC SMALL EXTRACELLULAR FLUID VOLUME. IT IS LIKELY THAT THIS ANIMAL WILL DEVELOP:
    A. Thirst
    B. Very rapid satiety (for water) with small volumes of water ingested
    C. A tendency to retain excessive water for more than 24 hours if it drinks a large volume
    D. Edema
    Ref. p. 422

354. AFTER DRINKING WATER:
    A. Thirst will not be relieved while the water stays in the gastrointestinal tract, unabsorbed
    B. Thirst will be relieved only after the ingested water enters the cells
    C. Thirst will be satisfied for a short time even if an esophageal fistula allows the water that entered the mouth to be lost again
    D. The extracellular water is increased significantly within 2 to 3 minutes of swallowing
    Ref. p. 423

355. THERE IS A "DRINKING CENTER" IN THE CENTRAL NERVOUS SYSTEM, AND IT IS LOCATED:
    A. In the nucleus of the glossopharyngeus nerve
    B. In the nucleus of the vagus
    C. Immediately adjacent to the supraoptic nuclei
    D. In the mammillary bodies
    E. In the pineal gland
    Ref. p. 423

356. THE INTERSTITIAL FLUID:
    A. Has a large mobile component, cushioning cells and serving as the true "cellular environment"
    B. Is mostly bound to mucopolysaccharides
    C. Is largely dependent on physiological fluctuations of the cardiovascular system and not on mucopolysaccharides
    D. Amounts to about 5% of the extracellular fluid space
    Ref. p. 424

357. A PERSON WEIGHING 65 KILOGRAMS WHO DRINKS ONE LITER OF ISOTONIC SALINE WILL UNDERGO ALL THE FOLLOWING PHYSIOLOGICAL CHANGES, EXCEPT:
    A. A third of one liter will remain in the cardiovascular system while two-thirds are confined to the interstitial space
    B. There is an increased cardiac output
    C. Blood pressure is increased
    D. Peripheral resistance decreases
    E. The kidneys remove the excess volume over a prolonged (24 hours) period of time
    Ref. p. 424

358. A PROLONGED INCREASE IN THE LEVEL OF ALDOSTERONE WILL RESULT IN ONLY A LIMITED EXPANSION OF THE EXTRACELLULAR FLUID VOLUME BECAUSE:
    A. An expanded ECF volume results eventually in an increased diuresis that overcomes the influence of aldosterone
    B. Aldosterone loses its effectiveness after a certain period of time
    C. Aldosterone inhibits the ADH secreting cells and only a limited volume of water can be reabsorbed
    D. All of the above
    Ref. p. 425

SECTION II - PHYSIOLOGY OF THE BODY FLUIDS 73
AND THE KIDNEYS

359. URINARY OUTPUT DECREASES IF:
  A. The carotid sinus baroreceptor is stimulated by a drop in blood pressure and responds by a reflex increase in arterial pressure
  B. The left atrium and other volume receptors are expanded by an increased blood volume
  C. The distal tubule is insensitive to antidiuretic hormone (genetically induced)
  D. Thirty per cent of the glomerular filtration enters the distal tubule
  Ref. p. 425

360. THE REASON WHY CARBONIC ACID DECREASES THE pH OF A WATER SOLUTION TO A VERY SMALL DEGREE IS THAT:
  A. Its bicarbonate ions have a very small affinity for hydrogen ions
  B. Its bicarbonate ions have a very intense affinity for hydrogen ions
  C. Water molecules in contact with bicarbonate ions dissociate H_3O^+ to a greater extent
  D. Bicarbonate ions in carbonic acid solutions are completely dissociated
  Ref. p. 428

361. A WATER SOLUTION OF BICARBONATE IONS OBTAINED BY INCREASING THE PRESSURE OF CARBON DIOXIDE (IN GAS FORM) IN THE LIQUID WILL:
  A. Have about 1000 times more bicarbonate ions than $H+$ ions
  B. Have about 1000 times more dissolved CO_2 molecules in gas form than bicarbonate ions in solution
  C. Have twice as many hydrogen ions than bicarbonate ions
  D. Not develop a pCO_2 (partial pressure of carbon dioxide) greater than 1/10 of an atmosphere due to the diffusibility of CO_2 from the liquid
  Ref. p. 429

362. IN A WATER SOLUTION CONTAINING CARBON DIOXIDE AND CARBONIC ACID IT IS TECHNICALLY EASIER TO MEASURE DIRECTLY THE:
  A. Amount of dissolved CO_2
  B. Concentration of undissociated carbonic acid
  C. Concentration of dissociated carbonic acid
  D. Concentration of hydronium (H_3O^+) ions
  Ref. p. 429

363. THE pH EQUALS THE pK IN A SOLUTION CONTAINING BICARBONATE AND CARBONIC ACID WHEN THE MOLAR CONCENTRATION OF BICARBONATE IS EQUAL TO:
  A. One-half the molar concentration of carbonic acid
  B. The molar concentration of carbonic acid
  C. Twice the molar concentration of carbonic acid
  D. Three times the molar concentration of carbonic acid
  Ref. p. 429

364. THE ADDITION OF SODIUM HYDROXIDE TO A SOLUTION CONTAINING CARBONIC ACID WILL CAUSE THE:
  A. Hydrogen ion concentration to increase
  B. Bicarbonate ion concentration to increase
  C. Sodium ion to become covalently bound to hydroxyl ions
  D. Removal of carbonic acid by an increased rate of evaporation
  Ref. pp. 429-430

SECTION II - PHYSIOLOGY OF THE BODY FLUIDS AND THE KIDNEYS

365. THE BUFFERING POWER OF A BICARBONATE:CARBONIC ACID SOLUTION IS GREATEST WHEN:
   A. A large fraction of bicarbonate ions is chemically balanced by sodium ions and only a small amount by hydrogen ions
   B. The pH equals the pK of the system
   C. The molar concentration of the buffer is small
   D. The pH of the solution is far removed from the point of inflection of the titration curve (acid added versus pH)
   Ref. p. 430

366. BICARBONATE IS A MORE IMPORTANT BUFFER THAN PHOSPHATE IN THE EXTRACELLULAR FLUIDS BECAUSE:
   A. Bicarbonate has its pK closer to the pH of the extracellular fluids
   B. Phosphate is not constantly renewed by ingestion
   C. Bicarbonate has a molar concentration many times that of phosphate
   D. Phosphate is not efficiently removed by the kidneys
   Ref. p. 431

367. IT IS VERY LIKELY TRUE THAT MOST OF THE ACIDS FORMED WITHIN THE BODY ARE BUFFERED BY:
   A. Intracellular protein
   B. Circulating hemoglobin
   C. The bicarbonate buffer system of the extracellular fluids
   D. The phosphate buffer system of the extracellular fluids
   Ref. p. 431

368. IN A NORMAL HUMAN AT REST, A SUDDEN TWO-FOLD INCREASE IN ALVEOLAR VENTILATION FOR ABOUT 15 MINUTES, WILL:
   A. Cause the pH in blood to be displaced to the alkaline side by approximately 0.2 units
   B. Cause acidosis due to the large proportion of bicarbonate blown off
   C. Not cause any pH displacement because other buffers will compensate for changes in bicarbonate
   D. Have no physiological effect whatsoever
   Ref. Fig. 37-3, p. 432

369. AN ACCUMULATION OF SODIUM BICARBONATE IN BLOOD WHICH IS CAPABLE OF RAISING THE pH TO ABOUT 7.6 WILL:
   A. Increase alveolar ventilation twofold
   B. Decrease alveolar ventilation to about 10% of its control value
   C. Suppress ventilation completely
   D. Cause a mild depression of the respiratory center with a mild decrease in alveolar ventilation to about 75% of its control value
   Ref. Fig. 37-4, p. 433

370. WHEN A NON-RESPIRATORY AILMENT CAUSES ACIDITY OF THE EXTRACELLULAR FLUID, THE RESPIRATORY COMPENSATION WILL BE:
   A. Complete, obtaining a pH of 7.4 in most cases
   B. Excessive leading the pH to the alkaline side of normal values
   C. Only partial
   D. Minimal
   Ref. p. 433

SECTION II - PHYSIOLOGY OF THE BODY FLUIDS AND THE KIDNEYS

371. CARBON DIOXIDE IN PLASMA OF PERITUBULAR CAPILLARIES OF THE KIDNEY CONTRIBUTES TO THE SECRETION OF HYDROGEN IONS BY:
    A. Combining with water
    B. Inhibiting carbonic anhydrase
    C. Removing hydrogen ions from ammonia
    D. Removing hydrogen ions from NaH_2PO_4
    Ref. p. 434

372. THE BULK OF HYDROGEN IONS SECRETED ALONG THE LENGTH OF RENAL NEPHRONS APPEARS IN THE TUBULAR FLUID AT THE LEVEL OF THE:
    A. Proximal tubule
    B. Loop of Henle
    C. Distal tubule
    D. Collecting duct
    Ref. p. 434

373. THE $\dfrac{\text{TUBULAR FLUID}}{\text{PLASMA}}$ HYDROGEN ION CONCENTRATION GRADIENT IS STEEPEST AT THE LEVEL OF THE:
    A. Proximal tubule
    B. Loop of Henle
    C. Distal tubule
    D. Collecting duct
    Ref. p. 434

374. THE MAXIMAL $\dfrac{\text{TUBULAR FLUID}}{\text{PLASMA}}$ HYDROGEN ION CONCENTRATION GRADIENT THAT CAN BE ACHIEVED REPRESENTS AN INCREASE EQUAL TO:
    A. 20 times the level in plasma
    B. 50 times the level in plasma
    C. 500 times the level in plasma
    D. 900 times the level in plasma
    E. 2000 times the level in plasma
    Ref. p. 434

375. THE RATE OF RENAL TUBULAR SECRETION OF HYDROGEN IONS IS DIRECTLY PROPORTIONAL TO THE:
    A. Amount of buffer in the tubular lumen
    B. Amount of sodium ions in the tubular lumen
    C. pH of the blood perfusing the renal tubules
    D. pCO_2 in blood perfusing the renal tubules
    Ref. p. 434

376. AN ALKALOSIS INDUCED BY AN EXCESS OF BICARBONATE IN THE EXTRACELLULAR FLUID CAUSES:
    A. The filtration of the tubular bicarbonate in excess and an increase in the tubular load of bicarbonate
    B. A complete titration of the tubular bicarbonate with an increased rate of secretion of hydrogen ions
    C. The reabsorption of the bulk of the filtered bicarbonate
    D. A more active reabsorption of sodium in the proximal and distal tubule with little sodium lost in the urine
    Ref. p. 435

SECTION II - PHYSIOLOGY OF THE BODY FLUIDS AND THE KIDNEYS

377. IN A CASE OF METABOLIC ACIDOSIS, AN INCREASED RATE OF SECRETION OF HYDROGEN IONS BY THE RENAL TUBULES CAUSES:
    A. An eventual increase in bicarbonate in the ECF
    B. A severe loss of sodium in the urine
    C. No real compensation to the acid-base disturbance
    D. An exacerbation of the acidosis   Ref. p. 435

378. AS DI-SODIUM PHOSPHATE IS CONVERTED INTO MONO-SODIUM PHOSPHATE IN THE RENAL TUBULAR FLUID, THE REACTION CONTRIBUTES TO:
    A. The increase of sodium bicarbonate concentration in the ECF
    B. Making the urine more alkaline
    C. Conserve hydrogen ions
    D. The excretion of sodium   Ref. p. 435

379. ALL OF THE FOLLOWING STATEMENTS ARE CORRECT, EXCEPT:
    A. Ammonia is synthesized not only in the proximal tubules but in the distal tubules and collecting ducts as well
    B. Ammonia diffuses in gaseous form out of the cells and into the tubular lumina
    C. Hydrogen ions convert ammonia into ammonium ions in the lumina of the tubules
    D. Ammonium ions are charged positively
    E. Ammonium ions are secreted in order to aid in the excretion of bicarbonate ions when the sodium concentration in plasma is low
           Ref. p. 436

380. AMMONIUM CHLORIDE DISSOLVED IN THE URINE:
    A. Has a very low degree of dissociation as it is a weak alkaline substance
    B Reaches a high concentration in prolonged alkalosis
    C. Reaches higher concentrations in prolonged metabolic acidosis
    D. Lowers the urine pH   Ref. p. 436

381. THE RENAL CAPACITY FOR ACID EXCRETION IN THE HUMAN:
    A. Has been estimated as being able to remove in the urine the hydrogen ions contained in one-half liter of 1.0 N hydrochloric acid in one day
    B. Is believed to be capable of excretion of very little alkali as compared to the large amount of excretable acid
    C. Permits a minimum pH of 2.0
    D. Permits a maximal urinary pH of 11.0
           Ref. p. 436

382. A SEVERE BOUT OF PNEUMONIA IS MOST LIKELY TO RESULT IN:
    A. Respiratory acidosis
    B. Respiratory alkalosis
    C. Metabolic acidosis
    D. Metabolic alkalosis   Ref. p. 438

SECTION II - PHYSIOLOGY OF THE BODY FLUIDS 77
AND THE KIDNEYS

383. A SEVERE CASE OF POLIOMYELITIS WITH INVOLVEMENT OF THE
MEDULLA OBLONGATA IS LIKELY TO LEAVE AS A CHRONIC AFTER-
EFFECT A MILD TO MODERATE CASE OF:
A. Respiratory acidosis
B. Respiratory alkalosis
C. Metabolic acidosis
D. Metabolic alkalosis   Ref. p. 438

384. SEVERE DIARRHEA TENDS TO CAUSE METABOLIC ACIDOSIS BECAUSE:
A. Intestinal secretions contain large amounts of bicarbonate
B. The infections causing diarrhea tend to form an excessive amount of acid
C. The gastric secretions containing hydrochloric acid are no longer neutralized by duodenal secretions at sufficiently high rates
D. The kidney tends to retain hydrogen ions when the body loses an excess of water   Ref. p. 438

385. PROLONGED AND EXCESSIVE VOMITING, AS IS THE CASE WITH IN-
TESTINAL OBSTRUCTIONS, WILL MOST LIKELY RESULT IN:
A. Respiratory acidosis
B. Respiratory alkalosis
C. Metabolic acidosis
D. Metabolic alkalosis   Ref. p. 438

386. VOMITING CAUSED BY TIGHT PYLORIC OBSTRUCTION SUCH AS
OCCURS IN CASES WITH SCARRED DUODENAL ULCER WITH PRE-
DOMINANT LOSS OF GASTRIC JUICE IS LIKELY TO RESULT IN:
A. Respiratory acidosis
B. Respiratory alkalosis
C. Metabolic acidosis
D. Metabolic alkalosis   Ref. p. 438

387. THE PATIENT SUFFERING A SEVERE CASE OF METABOLIC ACIDOSIS
DUE TO RETENTION OF FIXED ACIDS (DIABETES, UREMIA) WILL
SHOW:
A. Mental excitability
B. Depression of most of the central nervous system functions
C. Epilepsy
D. Hyperreflexia   Ref. p. 438

388. OVEREXCITABILITY OF THE CENTRAL NERVOUS SYSTEM SEEN IN
CASES OF ALKALOSIS IS NOT A UNIVERSAL FEATURE OF THE DYS-
FUNCTION OF THE NERVOUS SYSTEM. PORTIONS OF IT ARE DE-
PRESSED, AS IS THE CASE WITH THE:
A. Peripheral nerves
B. End-plates in neuromuscular junctions
C. Respiratory center
D. Cerebral cortex   Ref. p. 439

SECTION II - PHYSIOLOGY OF THE BODY FLUIDS AND THE KIDNEYS

389. PERSONS GENETICALLY PREDISPOSED TO DEVELOP CONVULSIONS (CONGENITAL EPILEPTICS) SHOULD BE ADVISED TO AVOID:
    A. Studying
    B. Acid producing foods
    C. Overbreathing
    D. Breath holding   Ref. p. 439

390. THE URINE OF A PERSON WHOSE RESPIRATORY DYSFUNCTION HAS CAUSED A CHRONIC RESPIRATORY ACIDOSIS (SEVERE CHRONIC EMPHYSEMA, FOR EXAMPLE) WILL MOST LIKELY CONTAIN AN EXCESS OF:
    A. Hydrogen ions
    B. Bicarbonate ions
    C. Potassium ions
    D. Sodium ions   Ref. p. 439

391. WHEN INTRAVENOUS THERAPY AIMED AT CORRECTING A CASE OF METABOLIC ACIDOSIS CONSISTS OF AN INFUSION OF SODIUM LACTATE, THE LACTATE ANION WILL, AFTER BEING METABOLIZED, BE REPLACED BY THE BODY'S NATURAL PROCESSES AND UNDER IDEAL CONDITIONS OF pH CORRECTION BY:
    A. Chloride
    B. Phosphate
    C. Gluconate
    D. Bicarbonate   Ref. p. 439

392. BLOOD pH, MEASURED UNDER CONDITIONS THAT DO NOT RESTRICT EVAPORATION OF THE NATURALLY DISSOLVED CARBON DIOXIDE, WILL CAUSE THE pH TO BE IN ERROR AS THERE WILL BE A:
    A. Shift to a more acid pH
    B. Shift of the pK of the carbonic acid/bicarbonate buffer
    C. Drop in the pO_2 which in turn will cause a change in total CO_2 content
    D. Shift to a more alkaline pH   Ref. p. 439

393. A PERSON WHOSE pH EQUALS 7.6 AND WHO HAS A PLASMA BICARBONATE CONCENTRATION EQUAL TO 38 mM/L PLASMA MAY, BY USING THE pH-BICARBONATE DIAGRAM, BE DIAGNOSED AS HAVING:
    A. An uncompensated metabolic alkalosis
    B. A metabolic alkalosis compensated by extra retention of carbonic acid
    C. A metabolic alkalosis complicated by a respiratory alkalosis
    D. A normal condition   Ref. Fig. 37-9, p. 440

394. IF THE PERSON DESCRIBED IN THE PREVIOUS QUESTION RETAINS THE SAME BICARBONATE CONCENTRATION IN PLASMA BUT DEVELOPS A pCO_2 10 mm HIGHER THAN BEFORE, THE pH IS MOST LIKELY:
    A. More alkaline than before
    B. Equal to 7.2
    C. Equal to 7.6
    D. Approaching compensation of the metabolic alkalosis
    E. Reflecting an imminent death   Ref. Fig. 37-9, p. 440

SECTION II - PHYSIOLOGY OF THE BODY FLUIDS AND THE KIDNEYS

395. A PERSON WHOSE pH EQUALS 7.2 AND WHOSE PLASMA BICARBONATE CONCENTRATION EQUALS 15 mM, HAS MOST LIKELY A(N):
    A. Uncompensated respiratory alkalosis
    B. Uncompensated respiratory acidosis
    C. Uncompensated metabolic alkalosis
    D. pCO_2 equals 40 mm Hg (1.2 mM CO_2 in the dissolved form)
    Ref. Fig. 37-9, p. 440

396. A PERSON WHOSE PLASMA VALUES WERE THOSE DESCRIBED IN THE PREVIOUS QUESTION WOULD BE MOST LIKELY A(N):
    A. Asthmatic
    B. Normal person holding his breath
    C. Diabetic
    D. Patient with gastric ulcer who ingested a large dose of sodium bicarbonate Ref. p. 438

397. PLASMA ANALYZED FOR BICARBONATE BY THE METHOD THAT DEFINES THE SO-CALLED "CARBON DIOXIDE COMBINING POWER":
    A. Gives totally useless results
    B. Gives data that allow the diagnosis of acid-base disturbance of respiratory origin
    C. Gives information on the CO_2 content equilibrated with a gas phase with 7% CO_2
    D. Allows the diagnosis of the acid-base disturbance without a pH measurement Ref. p. 440

398. THE DETRUSOR MUSCLE OF THE URINARY BLADDER UNDERGOES CONTRACTION DURING THE ACT OF MICTURITION. THIS IS STIMULATED BY IMPULSES CARRIED IN NERVE FIBERS THAT EMERGE:
    A. From the sacral segment of the spinal cord, as parasympathetic fibers
    B. As pudic nerves
    C. As sympathetic fibers from the lumbar segments of the spinal cord
    D. With the pelvic branches of the vagus nerves
    Ref. p. 442

399. THE INITIATION OF THE ACT OF MICTURITION INVOLVES THE:
    A. Contraction of the external sphincter of the bladder
    B. Contraction of the internal sphincter of the bladder
    C. Relaxation of the external sphincter of the bladder
    D. Contraction of the trigone Ref. p. 442

400. MICTURITION IS:
    A. A spinal cord reflex that can function without intervention of the brain when the lumbar portion of the spinal cord is compressed
    B. Not influenced by the brain in any way
    C. Caused by intravesical pressures equal to 10 mm of water under normal conditions
    D. Normally inhibited by descending pathways that originate in the lower levels of the medulla oblongata Ref. p. 444

SECTION II - PHYSIOLOGY OF THE BODY FLUIDS AND THE KIDNEYS

401. EXPOSURE OF A HUMAN TO TOXIC QUANTITIES OF CARBON TETRA-CHLORIDE WILL CAUSE LESIONS IN THE KIDNEY (ASIDE FROM OTHERS IN THE LIVER) WHICH RESULT IN A PREDOMINANT LOSS OF FUNCTION OF THE:
    A. Glomeruli
    B. Renal tubules
    C. Arterioles (afferent and efferent)
    D. Renin-Angiotensin system
    E. Pelvic-ureteral peristaltism   Ref. p. 445

402. AN INCOMPATIBLE TRANSFUSION CAUSES RENAL FUNCTION TO BE DISTURBED, DUE TO:
    A. Necrosis of renal vessels
    B. Abnormal quantities of hemoglobin products in the renal tubular lumina
    C. Calculi in the ureters
    D. A decrease in the oncotic pressure within glomerular capillaries
                                                Ref. p. 446

403. MYOGLOBIN:
    A. Unlike hemoglobin does not pass the glomerular capillary membrane
    B. May cause tubular necrosis through vasoconstriction and tubular obstruction
    C. Appears in plasma in cases of incompatible blood transfusion
    D. Appearing in blood plasma is never associated with shock due to its hypertensive properties   Ref. p. 446

404. THE FUNCTIONAL RESERVE OF BOTH KIDNEYS IS SO LARGE THAT RENAL INSUFFICIENCY WILL NOT REFLECT ITSELF IN ABNORMAL RETENTION OF METABOLIC PRODUCTS IN BLOOD BEFORE DAMAGE OF RENAL TISSUE HAS DESTROYED:
    A. One-tenth of the total renal parenchyma
    B. One-fourth of the total renal parenchyma
    C. Two-fourths of the total renal parenchyma
    D. Three-fourths of the total renal parenchyma
                                                Ref. p. 447

405. CHRONIC RENAL INSUFFICIENCY IS CHARACTERIZED BY:
    A. Small daily urine volumes equal to about 100-200 ml urine/24 hours
    B. Large daily urine volumes
    C. Concentrated urine (specific gravity 1.025)
    D. A small quantity of osmotic particles excreted per day even when the person ingests a large load of renal excretable material
                                                Ref. p. 447

406. ANEMIA FREQUENTLY FOUND IN CHRONIC RENAL FAILURE IS CONTRIBUTED BY THE:
    A. Insufficient amount of erythropoietin normally produced in the kidneys
    B. Lack of vitamin B_{12}
    C. Excessive loss of folic acid
    D. Insufficient retention of iron   Ref. p. 447

SECTION II - PHYSIOLOGY OF THE BODY FLUIDS
AND THE KIDNEYS

407. DEMINERALIZATION OF BONES IS A FREQUENT FINDING IN RENAL INSUFFICIENCY. IT IS BELIEVED THAT THIS IS CAUSED BY:
   A. The metabolic alkalosis frequently found in renal insufficiency
   B. The metabolic acidosis that frequently complicates renal insufficiency
   C. Atrophy of the parathyroid glands induced by uremia
   D. Low levels of plasma phosphate in this disease
   Ref. p. 448

408. THE REASON WHY THE HUMAN BODY SUBJECTED TO THE ACUTE LOSS OF RENAL FUNCTION HAS A TENDENCY TO DEVELOP ACIDOSIS LIES IN THE FACT THAT THE NORMAL BODY:
   A. Synthesizes an excess of 500 mM of acids over alkali in a day
   B. The kidney has more difficulty in excreting acids than in excreting alkali
   C. The respiratory centers respond to a rising concentration of hydrogen ions by a depressed ventilation which adds respiratory acidosis to the initial metabolic acidosis
   D. Sodium bicarbonate is rapidly lost by the insufficient kidney
   Ref. p. 448

409. POTASSIUM INTOXICATION:
   A. Is unavoidable in renal failure because cellular breakdown releases potassium ions to the extracellular fluid
   B. Can be averted indefinitely in a person deprived of both kidneys provided the person does not eat protein
   C. Will occur only after plasma potassium reaches a value equal to 20 mEq K+/L plasma
   D. Will occur even if the plasma potassium concentration reaches a value equal to 5 mEq/L
   Ref. p. 448

410. IN RENAL INSUFFICIENCY, THE INCREASE IN BLOOD LEVELS OF UREA:
   A. Reflects the predominant suppression of function of the tubules and not the absence of glomerular filtration
   B. Is damaging to the brain's metabolism even at relatively low levels (50 mg urea/100 ml)
   C. Is largely innocuous by itself
   D. Does not contribute to the renal osmotic load
   Ref. p. 448

411. THE FLUID USED IN AN "ARTIFICIAL KIDNEY" TO DIALYZE THE ABNORMAL PLASMA CONSTITUENTS OUT OF THE BLOOD, WHEN TREATING A UREMIC PATIENT, SHOULD CONTAIN:
   A. A lower sodium concentration than the plasma to be dialyzed
   B. No potassium ions at all
   C. A lactate ion concentration far higher than that in the patient's plasma to compensate osmotically the high concentration of urea in the patient's plasma
   D. A lower chloride concentration and a higher glucose concentration than the patient's plasma
   Ref. Table 38-1, p. 449

SECTION II - PHYSIOLOGY OF THE BODY FLUIDS AND THE KIDNEYS

412. IN A SEVERE CASE OF NEPHROSIS IT IS LIKELY THAT THE:
    A. Blood volume is large
    B. Extracellular fluid volume is small because of the excess fluid lost in the urine
    C. Renin-angiotensin-aldosterone mechanism is activated as a result of a reduced blood volume
    D. Oncotic pressure is normal (25 mm Hg)
    Ref. p. 450

413. A SIDE EFFECT OF MANY CASES OF NEPHROSIS IS A(N):
    A. Increased concentration of blood cholesterol
    B. Decrease in concentration of blood lipids
    C. Decrease of lipoprotein concentration in blood
    D. Mild dehydration
    Ref. p. 450

414. THE PERSON CONGENITALLY AFFLICTED BY THE SO-CALLED "RENAL TUBULAR ACIDOSIS" LOSES AN EXCESS OF:
    A. Hydrogen ions
    B. Bicarbonate ions
    C. Lactate ions
    D. Magnesium ions
    Ref. p. 451

415. PATIENTS WITH RENAL HYPOPHOSPHATEMIA:
    A. Have resistance to vitamin D therapy
    B. Never develop rickets
    C. Are very sensitive to minute fluctuations of phosphate in the diet
    D. Develop hypophosphatemia secondary to a selective retention of calcium ions by the kidney
    Ref. p. 451

416. THE INHIBITION OF CARBONIC ANHYDRASE WITHIN PROXIMAL TUBULAR CELLS CAUSED BY CERTAIN PHARMACOLOGICAL DIURETICS FACILITATES THE LOSS IN THE URINE OF:
    A. Hydrogen ions
    B. Bicarbonate ions
    C. Glucose molecules
    D. Glycine molecules
    Ref. p. 453

417. THE REASON WHY BEER CAUSES DIURESIS, WHEN INGESTED, LIES IN THE FACT THAT:
    A. Alcohol contained in the beer stimulates the secretion of antidiuretic hormone
    B. The water in the beer increases the glomerular filtration rate
    C. There is inhibition of tubular mechanisms of sodium reabsorption caused by alcohol
    D. Alcohol and water inhibit antidiuretic hormone release from the hypothalamus
    Ref. p. 453

SECTION III - PHYSIOLOGY OF THE RESPIRATORY SYSTEM

FOR EACH OF THE FOLLOWING MULTIPLE CHOICE QUESTIONS, SELECT THE ONE MOST APPROPRIATE ANSWER:

418. THE MUSCLE(S) WHICH SEEM TO CONTRIBUTE TO INSPIRATION TO A GREATER DEGREE THAN OTHER INSPIRATORY MUSCLES IN NORMAL QUIET BREATHING IS/ARE THE:
    A. Scapular elevators plus anterior serrati
    B. Scaleni
    C. Sternocleidomastoids
    D. Diaphragm
    E. External intercostals   Ref. p. 456

419. NORMAL QUIET EXPIRATION IS:
    A. An active process requiring expenditure of muscular energy
    B. A passive process simply consisting of relaxation of previously contracted inspiratory muscles
    C. Aided by voluntary movements of the abdominal muscles
    D. Significantly influenced by the contraction of the posterior inferior serrati   Ref. p. 456

420. PROBABLY THE MOST IMPORTANT FACTOR IN THE CREATION OF A NEGATIVE PRESSURE IN THE FLUID THAT SEPARATES THE PARIETAL FROM THE VISCERAL PLEURA IS THE:
    A. Cohesive force created by two wet surfaces juxtaposed against each other
    B. Constant reabsorption of fluid into capillaries of the visceral pleura
    C. Elastic recoil of the thoracic cavity displaced from its position of equilibrium
    D. Presence of surfactant in neighboring superficial alveoli
                                                Ref. p. 457

421. IT IS CURRENTLY BELIEVED THAT THE PULMONARY ALVEOLI OF SOME INFANTS BORN WITH WHAT IS KNOWN AS "HYALINE MEMBRANE DISEASE" OWE THEIR ABNORMALITY TO THE:
    A. Unusual elastic properties of the lung parenchyma in such infants
    B. Presence of fluid within the alveoli, which makes expansion much more difficult
    C. Absence of surfactant
    D. Lack of a negative pressure in the pleural space
                                                Ref. p. 458

422. LUNGS WHICH EASILY EXPAND WHEN THE INTRAPLEURAL PRESSURE DROPS BELOW THE ALVEOLAR PRESSURE:
    A. Have a low compliance
    B. Have a high compliance
    C. Owe their distensibility to a low level of surfactant
    D. Have a high collagen fibers ratio
                  elastic   Ref. p. 458

SECTION III - PHYSIOLOGY OF THE RESPIRATORY SYSTEM

423. COMPLIANCE CURVES (VOLUME IN ORDINATES; PRESSURE IN ABSCISSAE) OF LUNGS ALONE, SEPARATED FROM THE THORACIC CAVITY, WHEN COMPARED TO SIMILAR CURVES OBTAINED WHEN THE LUNGS ARE INSIDE THE THORACIC CAVITY:
A. Are distinctly flatter in the initial part of curves for lungs alone
B. Have a higher slope throughout the whole expansion of the lungs
C. Have a lesser slope throughout the whole expansion of the lungs
D. Have a shape that resembles a number eight
Ref. p. 459

424. THE THORACIC INTRAPLEURAL PRESSURE CAN BE MEASURED QUITE ACCURATELY BY FOLLOWING FLUCTUATIONS OCCURRING IN THE:
A. Abdominal cavity
B. Trachea
C. Gastric cavity
D. Esophageal lumen
E. Pharyngeal cavity
Ref. p. 459

425. A NORMAL RATIO OF $\frac{\text{VOLUME}}{\text{PRESSURE}}$ IN HUMANS EQUALS APPROXIMATELY:
A. 0.6 liter/2.0 cms H_2O pressure change
B. 0.1 liter/2.0 cms H_2O pressure change
C. 0.05 liter/2.0 cms H_2O pressure change
D. 1.0 liter/2.0 cms H_2O pressure change
Ref. p. 459

426. IN CASES OF FIBROTIC SCARRING OF THE PLEURA, FOLLOWING AN EXTENSIVE PLEURAL INFLAMMATION, THE COMPLIANCE IS LIKELY TO BE:
A. Unchanged
B. Diminished
C. Increased
D. Very much increased
Ref. p. 459

427. THE AREA CIRCUMSCRIBED BY THE COMPLIANCE LOOP IN A NORMAL INDIVIDUAL, WHEN COMPARED TO THAT OBTAINED IN A PERSON WHO SUFFERS EXTENSIVE PULMONARY FIBROSIS IS LIKELY TO BE:
A. Greater than the second
B. Less than the second
C. Essentially equal as the second
D. No longer an ovoid but a straight line
Ref. p. 459

428. SUCH AN AREA IS A REFLECTION OF (PROPORTIONAL TO) THE:
A. Elastic component of the alveoli
B. Work of breathing
C. Degree of distensibility of the elastic fibers
D. Reciprocal of the airway resistance
Ref. p. 460

429. THE VOLUME OF AIR INHALED AND EXHALED RHYTHMICALLY IN THE PROCESS OF BREATHING IS KNOWN AS:
A. Residual volume
B. Breathing volume
C. Tidal volume
D. Complementary volume
Ref. p. 460

SECTION III - PHYSIOLOGY OF THE RESPIRATORY SYSTEM

430. THE BASE LINE FROM WHICH ONE STARTS MEASURING THE INSPIRATORY RESERVE VOLUME IS:
    A. At the end of a normal expiration
    B. At the end of a normal inspiration
    C. At the end of a forced expiration
    D. None of the above   Ref. p. 461

431. THE FUNCTIONAL RESIDUAL CAPACITY INCLUDES:
    A. Residual volume plus tidal volume
    B. Tidal volume and expiratory reserve volume
    C. Tidal volume and inspiratory reserve volume
    D. Residual volume and expiratory reserve volume
                                        Ref. p. 461

432. AT THE SO-CALLED "RESTING EXPIRATORY LEVEL" THE VOLUME CONTAINED WITHIN THE LUNGS EQUALS THE:
    A. Expiratory reserve volume
    B. Residual volume
    C. Total lung capacity
    D. Functional residual capacity   Ref. p. 461

433. AN ABNORMALLY LARGE RESIDUAL VOLUME:
    A. Will prevent gas diffusion through the alveolar capillary membranes
    B. Will minimize the blood's oxygen and carbon doxide changes between the last expiration and the next inspiration
    C. Is associated with a more powerful expiratory capacity
    D. Is commonly associated with a small chest size
                                        Ref. p. 461

434. PERIODICAL MEASUREMENTS OF VITAL CAPACITY IN A PERSON WITH LEFT VENTRICULAR CONGESTIVE HEART FAILURE WILL SHOW THAT THE VITAL CAPACITY:
    A. Will increase in proportion with the dyspnea
    B. Decreases as a function of pulmonary congestion
    C. Does not give quantitative values of any use and should be replaced by periodical measurements of compliance
    D. Will show smaller changes than if there were an isolated right ventricular failure causing a diminution in the pulmonary circulation
                                        Ref. p. 461

435. BY SPIROMETRY ALONE IT IS IMPOSSIBLE TO ASSESS QUANTITATIVELY THE:
    A. Inspiratory reserve volume
    B. Expiratory reserve volume
    C. Functional residual capacity
    D. Vital capacity   Ref. p. 462

SECTION III - PHYSIOLOGY OF THE RESPIRATORY SYSTEM

436. THE FOLLOWING VALUES WERE OBTAINED IN THE PULMONARY
LABORATORY IN A NORMAL MALE:

Volume of Nitrogen expired (washout method)
after a normal expiration   = 2,340 ml
Per cent of Nitrogen in the alveolar air before the washout removal of
nitrogen   = 78%

ONE MAY CONCLUDE FROM THIS INFORMATION THAT THE
FUNCTIONAL RESIDUAL CAPACITY EQUALS:
A. 1000 ml
B. 2000 ml
C. 3000 ml
D. 4000 ml

Hint: $\dfrac{\text{vol } N_2}{\text{vol FRC}} = \dfrac{\% N_2}{100}$

Ref. p. 462

437. THE FOLLOWING GRAPH WAS OBTAINED ON A HUMAN WHO EXHALED
INTO A NITROGEN METER AFTER INHALING ONCE 100% OXYGEN;
TOTAL VOLUME EXHALED (TIDAL VOLUME) EQUALED 500 cc;
AREA A=40 cm^2; AREA B=60 cm^2:

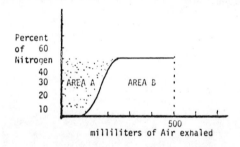

Redrawn with modifications from Fig.39-6
in Guyton's Textbook of Medical Physiology
W.B. Saunders

THE DEAD SPACE IN THIS CASE EQUALED:
A. 100 ml
B. 150 ml
C. 200 ml
D. 250 ml
E. 300 ml

Hint: $\dfrac{\text{Volume dead space}}{\text{Area A}} = \dfrac{\text{Tidal volume}}{\text{Area A + Area B}}$

Ref. Fig. 39-6, p. 463

SECTION III - PHYSIOLOGY OF THE RESPIRATORY SYSTEM

438. THE VOLUME KNOWN AS PHYSIOLOGICAL "DEAD SPACE":
   A. Does not change in any case as it is a fixed portion of the pulmonary space where no exchange of gases can occur
   B. Increases with exercise
   C. Increases when the person is in the supine position
   D. Averages about 40-50 ml of air in the normal adult male
   Ref. pp. 463-464

439. A PERSON WHOSE <u>TOTAL</u> PULMONARY VENTILATION EQUALS 2,000 ml/minute AND IS BREATHING AT A RATE OF 20 BREATHS PER MINUTE:
   A. Has a normal <u>alveolar</u> ventilation
   B. Has a normal tension of oxygen and carbon dioxide in arterial blood
   C. Can live normally provided he breathes a mixture of air with 20% oxygen
   D. Is in serious danger of death, as his ventilation is restricted to the dead space
   Ref. p. 464

440. THE FILTERING ACTION PROVIDED BY THE ANATOMICAL STRUCTURES OF THE NOSE AND UPPER RESPIRATORY PASSAGES ASSURES THAT THE MAXIMAL PARTICLE SIZE THAT REACHES THE ALVEOLI SUSPENDED IN THE AIR EQUALS:
   A. 1,000 microns
   B. 500 microns
   C. 10 microns
   D. 5 microns
   E. 0.5 microns
   Ref. p. 465

441. WHEN THE DEFENSIVE REFLEX OF COUGHING REACHES AN EXCESSIVELY HIGH FORCE, IT MAY CAUSE:
   A. The intraalveolar pressure to rise to a maximum of 10 mm Hg
   B. The intrapleural pressure to rise to +100 mm Hg
   C. The acceleration of tracheal air to maximal speeds equal to one centimeter per second (36 meters/hour)
   D. Collapse of the walls of the trachea and large bronchi
   Ref. p. 465

442. IN THE USE OF POSITIVE PRESSURE RESUSCITATORS THE PATIENT IS SUBJECTED TO:
   A. An increased cardiac output
   B. A decreased venous return during the positive pressure phase (inspiration)
   C. High blood pressure
   D. The same dangers encountered using hyperbaric chambers with pure oxygen at 2-3 atmospheres
   Ref. p. 468

443. THE FRACTION OF TOTAL PRESSURE CONTRIBUTED BY WATER VAPOR EVAPORATED FROM A THIN LAYER OF WATER IN CONTACT WITH A MIXTURE OF GASES AT 37° C:
   A. Depends on the volume of water present, being larger with larger volumes of the liquid
   B. Depends on the partial pressure of the other gases in the mixture
   C. Depends on the total pressure to which the mixture of gases is subjected
   D. Equals 47 mm Hg regardless of the partial pressure of the other gases
   Ref. p. 471

SECTION III - PHYSIOLOGY OF THE RESPIRATORY SYSTEM

444. THE SLOPE OF THE GRAPH THAT ILLUSTRATES THE FUNCTIONAL RELATIONSHIP BETWEEN MILLILITERS OF DISSOLVED GAS IN A UNIT VOLUME OF LIQUID (ORDINATES) AS A FUNCTION OF THE PRESSURE OF THE GAS IN THE GAS PHASE IN CONTACT WITH THE LIQUID IS LEAST IN THE CASE OF (SMALLEST SOLUBILITY COEFFICIENT):
    A. Oxygen
    B. Carbon dioxide
    C. Nitrogen
    D. Carbon monoxide   Ref. p. 427

445. THE SAME GRAPH REFERRED TO IN THE PREVIOUS QUESTION HAS A HIGHEST SLOPE IN THE CASE OF:
    A. Oxygen
    B. Carbon dioxide
    C. Nitrogen
    D. Carbon monoxide   Ref. p. 472

446. IF A PRESSURE OF 40 mm Hg IS USED TO DISSOLVE CARBON DIOXIDE IN WATER AT 37° C AND LATER NITROGEN (GAS) IS INTRODUCED AT A PRESSURE OF 600 mm Hg INTO THE GAS PHASE, THE VOLUME OF CARBON DIOXIDE ORIGINALLY DISSOLVED WILL:
    A. Decrease because of the presence of too many nitrogen molecules in the vicinity
    B. Increase because of an increased (600 + 40 mm Hg) total pressure
    C. Not change
    D. Be reduced to zero   Ref. p. 473

447. CARBON DIOXIDE DIFFUSES MORE EASILY THAN OXYGEN FROM BLOOD PLASMA THAT COMES IN CONTACT WITH THE ALVEOLAR CAPILLARY MEMBRANES IN THE LUNGS BECAUSE:
    A. Of its more favorable molecular weight
    B. The gradient between venous plasma and alveolar air is more favorable for carbon dioxide than for oxygen
    C. Of the greater mobility of carbon dioxide in the air space of the alveoli for a given pressure gradient between alveoli and the mouth
    D. Of the higher solubility coefficient for carbon dioxide than for oxygen
                                               Ref. p. 474

448. ALVEOLAR AIR COMPOSITION IS SUCH THAT:
    A. It can be totally freed of carbon dioxide during a very forceful and prolonged expiration or inspiration
    B. Its oxygen tension equals one-fifth of the nitrogen tension
    C. The pressure contributed by water vapor can decrease in a very dry desert (humidity of atmospheric air zero %) to a lesser value than the pCO_2
    D. Its oxygen fraction may be higher than that in expired air while the person breaths from a 100% oxygen source (tent, Douglas bag, etc.)
                                               Ref. p. 474

SECTION III - PHYSIOLOGY OF THE RESPIRATORY SYSTEM

449. IN REPLACING THE ALVEOLAR AIR BY AN ANESTHETIC MIXTURE ONE HAS TO KEEP IN MIND THE RATIO OF

$$\frac{\text{TIDAL AIR}}{\text{FUNCTIONAL RESIDUAL CAPACITY}}$$

AND THE RATE OF CHANGE OF COMPOSITION OF PULMONARY GASES. THUS, IN ORDER TO REACH AN EQUILIBRIUM WITH THE ANESTHETIC MIXTURE, AND ASSUMING A NORMAL ALVEOLAR VENTILATION (5 LITERS/MINUTE) THE ANESTHETIST MUST WAIT APPROXIMATELY:
A. 10-15 seconds
B. 20-30 seconds
C. 60-80 seconds
D. 5-6 minutes
E. 10-15 minutes
Ref. Fig. 40-6, p. 475

450. THE RATE OF OXYGEN CONSUMPTION AND THE ALVEOLAR VENTILATION ARE THE TWO FUNDAMENTAL FACTORS WHICH REGULATE THE PARTIAL PRESSURE OF OXYGEN IN THE ALVEOLAR SPACE. WHEN THE OXYGEN CONSUMPTION EQUALS 250 ML/MINUTE, AND THE ALVEOLAR VENTILATION EQUALS 5 LITERS/MINUTE, THE PARTIAL PRESSURE OF OXYGEN IN THE ALVEOLAR SPACE WILL BE MAINTAINED AT 110 mm Hg APPROXIMATELY. IF A PERSON ENGAGES IN MUSCULAR EXERCISE, AND HIS OXYGEN CONSUMPTION RISES TO 1,000 ml O_2/minute AND HIS ALVEOLAR VENTILATION EQUALS 10 L min, HIS PARTIAL PRESSURE OF OXYGEN IN THE ALVEOLAR SPACE WILL BE APPROXIMATELY EQUAL TO:
A. 110 mm Hg
B. 100 mm Hg
C. 90 mm Hg
D. 70 mm Hg
E. 10 mm Hg
Ref. p. 476

451. IN A CASE SIMILAR TO THAT DESCRIBED IN THE PREVIOUS QUESTION, THE PERSON WILL BE:
A. Euphoric
B. Perfectly comfortable and calm
C. Hypocapnic
D. Slightly anoxic
E. Dead
Ref. Fig. 40-7, p. 476

452. A NORMAL PERSON UNDERGOING THE TEST KNOWN AS MAXIMAL BREATHING CAPACITY IN A ROOM AT 760 mm Hg, BREATHING ROOM AIR, MAY THROUGH A MAXIMAL EFFORT INCREASE THE PARTIAL PRESSURE OF ALVEOLAR OXYGEN TO A VALUE NOT GREATER THAN:
A. 49 mm Hg
B. 99 mm Hg
C. 149 mm Hg
D. 209 mm Hg
E. 259 mm Hg

Hint: (760-47) x 0.21   Ref. Fig. 40-7, p. 476

SECTION III - PHYSIOLOGY OF THE RESPIRATORY SYSTEM

453. A PERSON ENGAGED IN MODERATELY SEVERE PHYSICAL EXERCISE WHO IS FORMING CARBON DIOXIDE AT A RATE EQUAL TO 800 ml PER MINUTE NEEDS TO HYPERVENTILATE IN ORDER TO MAINTAIN A NORMAL ALVEOLAR pCO_2. THIS INCREASED ALVEOLAR VENTILATION SHOULD REACH, IN ORDER TO MAINTAIN THE SAME pCO_2 (ABOUT 40 mm Hg), A VALUE APPROXIMATELY EQUAL TO:
A. 5 liters per minute
B. 10 liters per minute
C. 15 liters per minute
D. 20 liters per minute
E. 25 liters per minute
Ref. Fig. 40-8, p. 476

454. IF THE PERSON ALLEGED TO IN THE PREVIOUS QUESTIONS SHOULD HYPERVENTILATE AT A RATE EQUAL TO 30 L/MINUTE, HIS ALVEOLAR pCO_2 IS LIKELY TO BE EQUAL TO:
A. 5 mm Hg
B. 20 mm Hg
C. 30 mm Hg
D. 40 mm Hg
E. 50 mm Hg
Ref. Fig. 40-8, p. 476

455. SUCH A PERSON, AS DESCRIBED IN THE QUESTIONS ABOVE, WILL BE:
A. In perfectly normal condition and capable to continue his exercise for at least 15 minutes
B. In respiratory alkalosis
C. Extremely hypoxemic
D. Hypercapnic (excessive CO_2 retention)
Ref. Fig. 40-8, p. 476; and p. 438

456. THE SURFACE AREA OF THE ALVEOLAR-CAPILLARY MEMBRANES OVER WHICH GAS EXCHANGES OCCUR DURING RESPIRATION MAY BE COMPARED TO THE SURFACE AREA OF THE SKIN. IF THE LATTER EQUALS 2 M^2 (SQUARE METERS) IN A HUMAN ADULT, THAT OF THE LUNGS EQUALS APPROXIMATELY:
A. The same area as that of the skin
B. 10 times the area of the skin
C. 20 times the area of the skin
D. 40 times the area of the skin
E. 100 times the area of the skin
Ref. p. 477

457. THE VOLUME OF BLOOD THAT IS THINLY SPREAD OVER SUCH AREA AMOUNTS TO APPROXIMATELY:
A. 10-20 ml
B. 60-100 ml
C. 500-1000 ml
D. 3000-4000 ml
Ref. p. 477

458. NET DIFFUSION OF CARBON DIOXIDE FROM VENOUS BLOOD TO ALVEOLAR SPACES OCCURS BECAUSE:
A. The diffusibility of carbon dioxide in plasma water is less than in the gaseous phase of the alveoli
B. Of its large kinetic activity
C. The alveolar pCO_2 is lower than that in venous blood
D. Of its very light molecular weight
Ref. pp. 478-479

SECTION III - PHYSIOLOGY OF THE RESPIRATORY SYSTEM 91

459. ASSUMING AN OXYGEN CONSUMPTION EQUAL TO 240 ml/min, A MEAN (INTEGRATED) pO_2 IN THE ALVEOLAR CAPILLARY BLOOD EQUAL TO 88 mm Hg, AN ALVEOLAR pCO_2 EQUAL TO 100 mm Hg, THE SO-CALLED "DIFFUSING CAPACITY" FOR OXYGEN EQUALS:
   A. 10 ml O_2/mm Hg pO_2 gradient/min
   B. 20 ml O_2/mm Hg pO_2 gradient/min
   C. 30 ml O_2/mm Hg pO_2 gradient/min
   D. 40 ml O_2/mm Hg pO_2 gradient/min
   Ref. p. 479

460. UNDER NORMAL CONDITIONS, AVERAGE VALUES OF A pCO_2 GRADIENT (pCO_2 IN PULMONARY VEINS - pCO_2 IN ALVEOLAR AIR) AMOUNTS TO APPROXIMATELY:
   A. 1 mm Hg
   B. 10 mms Hg
   C. 20 mms Hg
   D. 50 mms Hg
   Ref. p. 479

461. IF THE RATIO DIFFUSING CAPACITY CO_2 HAD THE SAME VALUE AS DIFFUSING CAPACITY O_2
   THE RATIO DIFFUSING COEFFICIENT FOR CO_2 THE DIFFUSING DIFFUSING COEFFICIENT FOR O_2
   CAPACITY FOR CARBON DIOXIDE WOULD BE EQUAL TO:
   A. 20 ml CO_2/mm Hg pressure gradient/minute
   B. 50 ml CO_2/mm Hg pressure gradient/minute
   C. 100 ml CO_2/mm Hg pressure gradient/minute
   D. 200 ml CO_2/mm Hg pressure gradient/minute
   E. 400 ml CO_2/mm Hg pressure gradient/minute
   Ref. p. 479

462. A PATIENT SUFFERING FROM PROGRESSIVE RESPIRATORY FAILURE (SEVERE EMPHYSEMA) WITH PROGRESSIVE DIMINUTION OF THE ALVEOLAR CAPILLARY MEMBRANE SURFACE AREA WILL, IF NOT TREATED, DIE BECAUSE OF:
   A. Retention of carbon dioxide
   B. Hypoxia
   C. Retention of nitrogen
   D. Pulmonary edema
   Ref. p. 479

463. THE SINGLE BREATH CARBON MONOXIDE TEST FOR DIFFUSING CAPACITY:
   A. May be occasionally lethal
   B. Takes advantage of the fact that the partial pressure of carbon monoxide in plasma water is negligible because carbon monoxide is almost totally bound to hemoglobin
   C. Is used because of the higher diffusibility of carbon monoxide than oxygen across the alveolar capillary membrane
   D. Gives values equal to about 100 ml carbon monoxide diffusing per minute per millimeter of pCO gradient
   Ref. p. 480

SECTION III - PHYSIOLOGY OF THE RESPIRATORY SYSTEM

464. IF HUMANS COULD BE DEPRIVED OF THEIR RED BLOOD CELLS AND WERE TO DEPEND ONLY ON DISSOLVED OXYGEN TO SATISFY TISSUE NEEDS OF OXYGEN, THE CARDIAC OUTPUT OF A HEMOGLOBIN-FREE FLUID SHOULD BE HIGHER THAN THE NORMAL (OF 5 LITERS BLOOD/MINUTE) BY A FACTOR OF:
    A. 2-3 times
    B. 10-15 times
    C. 30-100 times
    D. 1000-2000 times   Ref. p. 481

465. AS BLOOD ENTERS THE PULMONARY CAPILLARIES THE DIFFUSION OF OXYGEN AND SATURATION OF HEMOGLOBIN PROCEEDS RAPIDLY. THUS, IN A PERSON BREATHING AIR AT 760 mm Hg, A SMALL VOLUME OF BLOOD CONTAINED IN A PULMONARY CAPILLARY MAY BE ESTIMATED TO HAVE OVER 90% OF ITS OXYGEN "LOADING" COMPLETED:
    A. In one-tenth of the time it spends in the capillary
    B. In two-tenths of the time it spends in the capillary
    C. In one-half of the time it spends in the capillary
    D. By the time it leaves the pulmonary capillary
                                                Ref. p. 482

466. THE ADMIXTURE OF A SMALL VOLUME OF VENOUS BLOOD WITH A LARGE VOLUME OF FULLY OXYGENATED BLOOD, DUE TO NORMAL SHUNTS IN THE PULMONARY CIRCULATION, MAY FREQUENTLY CAUSE:
    A. A drop in the pO_2 in arterialized blood equal to 1 mm Hg(from 99 mm to 98 mm Hg) with a large diminution in the oxygen content (from 20 to 17 volumes O_2 per 100 ml blood)
    B. The pO_2 in arterialized blood to drop 10 mm Hg (from 99 mm to 89 mm Hg) with a small diminution in oxygen content (from 20 to 19.8 volumes O_2 per 100 ml blood)
    C. Cyanosis
    D. A slight compensatory increase in pCO_2
                                                Ref. p. 482

467. BLOOD WHICH HAS FINISHED PERFUSING A TISSUE AND STARTS ITS RETURN TO THE HEART IS MOST LIKELY TO HAVE A:
    A. pO_2 similar to that in tissues
    B. pO_2 higher than that in tissues
    C. pCO_2 higher than that in tissues
    D. pO_2 lower than that in tissues   Ref. p. 483

468. UNDER NORMAL CONDITIONS, CELLULAR METABOLISM REQUIRES THAT THE OXYGEN WITHIN THE CELLS HAVE A TENSION NOT LESS THAN:
    A. 1-5 mm Hg
    B. 10-15 mm Hg
    C. 20-30 mm Hg
    D. 40-50 mm Hg   Ref. p. 483

SECTION III - PHYSIOLOGY OF THE RESPIRATORY SYSTEM 93

469. BECAUSE OF ITS SOLUBILITY AND OTHER PHYSICAL FEATURES THE INTERSTITIAL FLUID PARTIAL PRESSURE OF CARBON DIOXIDE HAS VALUES EQUAL TO:
A. 5-10 mms higher than the pCO_2 within cells
B. 15-20 mms higher than the pCO_2 within cells
C. 1-2 mms Hg lower than the pCO_2 within cells
D. 10-15 mms Hg lower than the pCO_2 within cells
Ref. p. 484

470. A VERY LARGE INCREASE IN BLOOD FLOW THROUGH A TISSUE AT REST, WHERE THE pCO_2 IS MEASURED AS BEING 40 mm Hg, WILL CAUSE THE VENOUS BLOOD pCO_2 TO DROP AS LOW AS APPROXIMATELY:
A. 10 mm Hg
B. 20 mm Hg
C. 30 mm Hg
D. 40 mm Hg
E. 50 mm Hg
Ref. p. 484

471. A SATURATED BLOOD SAMPLE OBTAINED FROM A POLYCYTHEMIC PATIENT WAS ANALYZED BY THE VAN SLYKE METHOD FOR OXYGEN CONTENT. IT CONTAINED 27.1 ml O_2/100 ml BLOOD. USING THE SOLUBILITY COEFFICIENT FOR OXYGEN (0.024 ml O_2/ml blood/mm Hg) IT WAS ESTIMATED THAT THERE WERE 0.3 ml OF OXYGEN/100 ml BLOOD IN THE DISSOLVED FORM, AND THAT 26.8 ml O_2/100 ml BLOOD WERE BOUND TO HEMOGLOBIN. KNOWING THAT THERE IS A CONSTANT VOLUME OF OXYGEN THAT CAN BE CARRIED BY EACH GRAM OF HEMOGLOBIN, ONE MAY SAY THAT IN THIS CASE THE HEMOGLOBIN CONCENTRATION WAS EQUAL TO:
A. 5 grams Hb/100 ml blood
B. 10 grams Hb/100 ml blood
C. 20 grams Hb/100 ml blood
D. 30 grams Hb/100 ml blood

Hint: $\dfrac{26.8 \text{ ml } O_2}{x \text{ grams Hb}} = \dfrac{1.34 \text{ ml } O_2}{1.0 \text{ gram Hb}}$ Ref. p. 485

472. THE USUAL DROP IN PARTIAL PRESSURE OF OXYGEN IN ARTERIAL BLOOD AS IT BECOMES VENOUS, IS OF THE ORDER OF 60 mm Hg (DROP FROM 100 mm Hg TO ABOUT 40 mms Hg). AS A CONSEQUENCE THE HEMOGLOBIN SATURATION, WHICH DEPENDS ON THE PRESSURE, DROPS TO VALUES EQUAL TO (FROM 98% SATURATION):
A. 5%
B. 25%
C. 50%
D. 70%
E. 80%
Ref. Fig. 41-8, p. 485

473. A DIMINUTION IN pO_2 AND IN PER CENT SATURATION SIMILAR TO THAT DESCRIBED IN THE PREVIOUS QUESTION MEANS THAT FOR EACH MILLILITER OF BLOOD PERFUSING THE VARIOUS ORGANS THE VOLUME OF OXYGEN DELIVERED EQUALS IN A HUMAN WITH 0.20 MILLILITERS OF OXYGEN/MILLILITER OF BLOOD:
A. 0.01 ml O_2/ml blood perfusing
B. 0.05 ml O_2/ml blood perfusing
C. 0.10 ml O_2/ml blood perfusing
D. 0.15 ml O_2/ml blood perfusing
Ref. p. 485

SECTION III - PHYSIOLOGY OF THE RESPIRATORY SYSTEM

474. WHEN TISSUE HYPOXIA BECOMES SEVERE AND TISSUE pO_2 IS DIMINISHED TO 8 mms Hg (INSTEAD OF THE USUAL 40 mm Hg), THE CORRESPONDING OXYGEN CONTENT IN BLOOD MAY DROP FROM 0.20 ml O_2 ml ARTERIAL BLOOD TO:
    A. 0.18 ml O_2 ml venous blood
    B. 0.15 ml O_2 ml venous blood
    C. 0.10 ml O_2 ml venous blood
    D. 0.07 ml O_2 ml venous blood
    E. 0.02 ml O_2 ml venous blood   Ref. Fig. 41-9, p. 485

475. THE 15-FOLD INCREASE IN THE RATE OF DELIVERY OF OXYGEN TO THE TISSUES DURING MUSCULAR EXERCISE IS MOST LIKELY DUE TO A:
    A. 15-fold increase in the cardiac output alone
    B. 15-fold increase in the oxygen utilization coefficient
    C. 3-fold increase in the oxygen utilization coefficient and a 5-fold increase in cardiac output
    D. 1-fold increase in the oxygen utilization coefficient and a 14-fold increase in cardiac output   Ref. p. 486

476. BY SUBJECTING A HUMAN BODY TO 4 ATMOSPHERES OF OXYGEN PRESSURE, THE HEMOGLOBIN DOES NOT SIGNIFICANTLY INCREASE ITS OXYGEN CONTENT ABOVE VALUES FOUND UNDER NORMAL CONDITIONS AT 100 mm Hg. THE PLASMA, ON THE OTHER HAND, MAY CARRY AS MUCH AS 8 ml O_2 100 ml PLASMA (WHICH COMPARES FAVORABLY TO 5 ml O_2 DELIVERED AT REST BY 100 ml OF NORMAL BLOOD AT 1 ATMOSPHERE). AT SUCH HYPERBARIC CONDITIONS, A SERIOUS CAUSE OF DANGER IS THE:
    A. Depression of the carbon dioxide transport
    B. Increase in the oxygen utilization
    C. Suppression of the hemoglobin buffering capacity
    D. High pressure effects on cardiac contractility
             Ref. p. 489

477. THE OXYGEN DISSOCIATION CURVE IS OBTAINED BY PLOTTING THE PER CENT OF HEMOGLOBIN WHICH IS (REVERSIBLY) BOUND TO OXYGEN AT PROGRESSIVELY INCREASING PRESSURES OF OXYGEN IN THE GAS PHASE IN CONTACT WITH BLOOD. VALUES HAVE BEEN OBTAINED AS FOLLOWS:

pO_2	Per cent saturation
10 mm Hg	10%
20 mm Hg	28%
40 mm Hg	66%
60 mm Hg	83%
80 mm Hg	92%

    WHEN OXYGEN IS COMPLETELY REPLACED BY CARBON MONOXIDE IN THE GAS PHASE IT IS ESTIMATED THAT 90% OF THE HEMOGLOBIN NOW BECOMES SATURATED WITH CARBON MONOXIDE AT PRESSURES EQUAL TO THOSE SHOWN BY LETTER:
    A. 30 mm Hg pCO
    B. 15 mm Hg pCO
    C. 3 mm Hg pCO
    D. 0.3 mm Hg pCO   Ref. Fig. 41-14, p. 490

SECTION III - PHYSIOLOGY OF THE RESPIRATORY SYSTEM

478. IF IN AN ENCLOSED ROOM A PERSON IS BREATHING A MIXTURE WHERE THE pO_2 EQUALS 90 mm Hg AND THE pCO (PARTIAL PRESSURE OF CARBON MONOXIDE) EQUALS 1 mm Hg, THE PERSON WILL MOST LIKELY:
A. Have euphoria
B. Experience no mental changes
C. Have hyperpnea
D. Be dead   Ref. p. 490

479. OF THE FOLLOWING MIXTURES, THE MOST APPROPRIATE TO REVIVE A PATIENT SUFFERING FROM CARBON MONOXIDE POISONING IS THE ONE CONATINING:
A. 20% oxygen, 5% carbon dioxide and 75% nitrogen
B. 50% oxygen, 5% carbon dioxide and 45% nitrogen
C. 60% oxygen, 5% carbon dioxide and 35% nitrogen
D. 95% oxygen, 5% carbon dioxide and no nitrogen
E. 99% oxygen, 1% carbon dioxide and no nitrogen
                                              Ref. p. 490

480. THE TRANSFORMATION OF CARBON DIOXIDE PRODUCED WITHIN CELLS IN GASEOUS FORM INTO BICARBONATE IONS TAKES PLACE MAINLY:
A. In plasma
B. Aided by carbonic anhydrase within erythrocytes
C. Due to the higher pCO_2 in venous blood
D. When the pH of venous blood is approximately equal to 7.6
                                              Ref. p. 490

481. THE PORTION OF TOTAL CARBON DIOXIDE REMOVED FROM TISSUES BY VENOUS BLOOD IN SOLUTION IN PLASMA WATER IS A PART OF THE TOTAL CARBON DIOXIDE TRANSPORTED FROM TISSUES TO LUNGS. THE PERCENTAGE TRANSPORTED IN DISSOLVED GASEOUS FORM IN PLASMA WATER EQUALS ABOUT:
A. 0.7% of the total
B. 7.0% of the total
C. 37% of the total
D. 67% of the total   Ref. p. 490

482. HYDROGEN IONS DISSOCIATED FROM CARBONIC ACID IN VENOUS BLOOD:
A. Are quickly buffered by hemoglobin molecules
B. Are mainly buffered by bicarbonate ions in plasma
C. Largely remain free and cause a significant pH shift to the acid side
D. Largely diffuse out of red cells and are buffered by albumin and other plasma protein molecules   Ref. p. 490

483. BICARBONATE IONS FORMED WITHIN RED CELLS:
A. Are carried in part as carbamino compounds in venous blood
B. Diffuse out of the cells and are carried in plasma
C. Are carried in bulk within erythrocyte membrane
D. Form a negligible part of the total carbon dioxide transported by venous blood to the lungs when compared to the carbamino fraction
                                              Ref. p. 491

SECTION III - PHYSIOLOGY OF THE RESPIRATORY SYSTEM

484. THE ADMINISTRATION OF A CARBONIC ANHYDRASE INHIBITOR WILL RESULT IN A(N):
    A. Increase in the concentration of bicarbonate in plasma
    B. Increase in the total amount of carbonic acid in plasma
    C. Low pCO_2 in tissues and a high pCO_2 in the alveoli
    D. High pCO_2 in tissues   Ref. p. 491

485. THE AMOUNT OF "CARBAMINO" COMPOUNDS IN BLOOD:
    A. Is about evenly distributed between erythrocytes and plasma
    B. Is largely present in plasma
    C. Is far more abundant within red blood cells
    D. Represents the bulk of carbon dioxide transported from tissues to lungs   Ref. p. 491

486. BLOOD PASSING THROUGH THE LUNGS IS EXPOSED TO A LOWER pCO_2 THAN THAT WHICH IN THE TISSUES INCREASED THE CARBON DIOXIDE CONTENT OF BLOOD. THE VOLUME OF CARBON DIOXIDE LOST IN THE LUNGS AS A RESULT OF THIS PRESSURE GRADIENT IN PULMONARY ALVEOLI EQUALS APPROXIMATELY:
    A. 0.5 milliliters of CO_2 / 100 ml blood
    B. 1.0 milliliters of CO_2 / 100 ml blood
    C. 5.0 milliliters of CO_2 / 100 ml blood
    D. 25.0 milliliters of CO_2 / 100 ml blood
    Ref. p. 491

487. WHEN BLOOD GAINS OXYGEN IN THE PULMONARY CAPILLARIES THE:
    A. Affinity of the hemoglobin molecule for carbon dioxide decreases
    B. Carbon dioxide dissociation curve is "shifted to the left"
    C. Hemoglobin molecule becomes more alkaline as hydrogen ions are withdrawn from solution
    D. Fraction of carbamino compound within the erythrocyte increases
    Ref. p. 492

488. WHEN BLOOD INCREASES ITS CARBON DIOXIDE CONTENT IN PERIPHERAL TISSUES:
    A. Red blood cells increase only the fraction of carbon dioxide carried as bicarbonate in plasma but not the carbon dioxide carried as carbamino compounds
    B. The affinity of hemoglobin for oxygen decreases moderately
    C. The hemoglobin molecule develops a lower tendency to dissociate hydrogen ions
    D. The oxygen dissociation curve is shifted to the left
    Ref. p. 492

489. THE RATIO $\dfrac{CO_2 \text{ PRODUCTION}}{O_2 \text{ CONSUMPTION}}$:
    A. Becomes less than 1.0 only when the person is at total rest
    B. Depends exclusively on the rate of ventilation at rest
    C. Varies proportionately with the cardiac output
    D. Is determined by the protein/fat/carbohydrate ratio metabolized by cells   Ref. p. 493

SECTION III - PHYSIOLOGY OF THE RESPIRATORY SYSTEM

490. THE RESPIRATORY CENTER IS:
   A. Restricted to the medulla oblongata and has insignificant facilitatory influence from suprasegmental levels
   B. Stimulated by the so-called "pneumotaxic" center
   C. Inhibited by fibers that reach the medulla oblongata with the hypoglossal nerve
   D. Endowed with an intrinsic oscillating rhythm
   Ref. p. 494

491. THE PATTERN OF PROLONGED AND DEEP INSPIRATIONS IS FACILITATED BY ACTIVITY OF THE SO-CALLED "APNEUSTIC CENTER" LOCATED IN THE:
   A. Medulla oblongata, separated from rostral levels
   B. Pons, separated from rostral levels
   C. Cerebral peduncles, separated from rostral levels
   D. Substantia nigra
   E. Cerebellum
   Ref. pp. 494-495

492. THE VAGUS NERVE TRANSMITS IMPULSES WHICH ACT:
   A. By facilitating the apneustic center
   B. In a manner similar to the pneumotaxic center
   C. To a very limited degree to modify respirations
   D. To convey efferent impulses to the diaphragm
   Ref. p. 496

493. FACILITATION OF THE ACTIVITY OF INSPIRATORY NEURONS COMES FROM ALL THE FOLLOWING SOURCES, EXCEPT:
   A. Spinal cord
   B. Apneustic center
   C. Pneumotaxic center
   D. Glossopharyngeus nerve
   E. Chemorreceptor areas in the carotid and aortic bodies
   Ref. p. 496

494. THE AREA OF THE CENTRAL NERVOUS SYSTEM EXQUISITELY SENSITIVE TO INCREASES IN HYDROGEN ION CONCENTRATION IN THE CEREBROSPINAL FLUID IS LOCATED AT THE:
   A. Outer surface of the medulla oblongata (level of the VIII and IX nerves)
   B. Outer surface of the pons (level of the V nerve)
   C. Cervical spinal cord (level of the XII nerve)
   D. Hypothalamus (mamillary bodies)
   Ref. pp. 497-498

495. INCREMENTS IN HYDROGEN ION CONCENTRATION THAT SEEM TO ACTIVATE NEURONS IN THE CHEMOSENSITIVE AREAS ARE BELIEVED TO BE ORIGINATED MAINLY IN:
   A. Larger concentrations of lactic acid in the cerebrospinal fluid
   B. Diminution of phosphate buffer and bicarbonate buffer in the cerebrospinal fluid
   C. Increases in the pCO_2 in the blood that perfuses the medulla oblongata
   D. Decreased blood flow through the brain
   Ref. p. 498

SECTION III - PHYSIOLOGY OF THE RESPIRATORY SYSTEM

496. IMPULSES ORIGINATED IN THE CAROTID CHEMORECEPTORS ASCEND TO THE CENTRAL NERVOUS SYSTEM WITH FIBERS OF THE:
    A. Glossopharyngeus nerve
    B. Vagus nerve
    C. Hypoglossal nerve
    D. Intermedius nerve
    E. Phrenic nerve
    Ref. p. 498

497. THE FUNCTIONAL RELATIONSHIP BETWEEN ALVEOLAR VENTILATION (PLOTTED AS ORDINATES) AND pCO_2 IN THE ALVEOLAR SPACE (PLOTTED AS ABSCISSAE) IS A LINEAR ONE. THE SLOPE OF THIS LINE IS:
    A. A negative one
    B. Increased by lower pO_2 in the alveoli
    C. Unchanged by pH
    D. Exponential
    Ref. Fig. 42-5, p. 499

498. AN ACUTELY DEVELOPED, MODERATELY SEVERE DECREASE IN OXYGEN CONCENTRATION IN THE ALVEOLAR AIR, THAT LASTS FOR SEVERAL HOURS, WILL MAKE THE SENSITIVITY OF THE RESPIRATORY CENTER TO CONCOMITANT INCREMENTS IN pCO_2 BECOME:
    A. Increased
    B. Decreased
    C. The same
    D. Totally suppressed
    Ref. p. 499 and Fig. 42-5

499. RESPONSIVENESS OF THE MEDULLARY CENTER IS EXPLORED IN HUMANS BY ALLOWING THEM TO BREATHE A 7.0-7.5% CO_2 MIXTURE IN OXYGEN AND MEASURING THE HYPERVENTILATORY RESPONSE. THE AVERAGE RESPONSE OBTAINED IN A GROUP OF NORMALS WHEN COMPARED WITH THE RESPONSE OBTAINED IN PATIENTS WITH METABOLIC ACIDOSIS (DIABETIC, UREMIC) SHOULD BE EXPECTED TO BE IN THE CASE OF THE LATTER GROUP:
    A. Less than the normals
    B. Greater than the normals
    C. Equal as the normals
    D. Greatly variable
    Ref. Fig. 42-5, p. 499

500. THE MAXIMAL CONCENTRATION OF CARBON DIOXIDE IN INSPIRED AIR THAT COULD CAUSE HYPERVENTILATION, AND BEYOND WHICH ANY FURTHER INCREMENT IN CARBON DIOXIDE CONCENTRATION WOULD START TO CAUSE PROGRESSIVE DEPRESSION OF THE CENTRAL NERVOUS SYSTEM IS:
    A. 5%
    B. 9%
    C. 15%
    D. 20%
    Ref. p. 500

SECTION III - PHYSIOLOGY OF THE RESPIRATORY SYSTEM 99

501. IF AN ANIMAL IS ENCLOSED IN A CHAMBER WHERE THE CARBON DIOXIDE IS REMOVED CONSTANTLY, AND THE ALVEOLAR pO_2 IS ALLOWED TO DROP FROM 100 mm Hg TO 40 mm Hg THERE WILL BE AN INCREASE IN ALVEOLAR VENTILATION. THE RATIO
$$\frac{\text{ALVEOLAR VENTILATION AT } pO_2 \text{ 40 mms Hg}}{\text{ALVEOLAR VENTILATION AT } pO_2 \text{ 100 mms Hg}}$$
WILL BE EQUAL UNDER THSE CONDITIONS TO APPROXIMATELY:
A. 1.2
B. 2.2
C. 5.2
D. 10.2
Ref. Fig. 42-6, p. 500

502. THE PHYSIOLOGICAL RESPONSE ELICITED BY THE PREVIOUS HYPOXIA WILL CAUSE THE ALVEOLAR pCO_2 TO:
A. Drop significantly
B. Rise at least 10 mms Hg
C. Stay at the same level
D. Rise at least 30 mms Hg
Ref. p. 501

503. IF THE ALVEOLAR pCO_2 IN THE EXPERIMENT QUOTED IN QUESTION #501 COULD BE MAINTAINED CONSTANT (INDUCING HYPOXIA BUT PREVENTING THE DROP IN ALVEOLAR pCO_2) THE HYPERVENTILATORY RESPONSE WILL BE DIFFERENT. THUS THE RATIO
$$\frac{\text{ALVEOLAR VENTILATION AT } pO_2 \text{ 40 mms Hg, } pCO_2 \text{ 40 mms Hg}}{\text{ALVEOLAR VENTILATION AT } pO_2 \text{ 100 mms Hg, } pCO_2 \text{ 40 mms Hg}}$$
WILL BE EQUAL TO ABOUT:
A. 0.2-0.5
B. 0.5-1.0
C. 1.0-1.5
D. 2.0-3.0
Ref. Fig. 42-6, Fig. 42-2, p. 497 and p. 500

504. A VARIATION OF ALVEOLAR VENTILATION BETWEEN HALF THE NORMAL VALUE AND TEN TIMES THE NORMAL VALUE HAS A MOST SIGNIFICANT EFFECT ON THE:
A. Rate of transport and delivery of oxygen by arterial blood to the tissues
B. Rate of transport and removal of carbon dioxide from tissues to the lungs
C. Affinity of oxygen and tissues
D. Rate at which hemoglobin binds to oxygen in alveolar capillaries
Ref. p. 501

505. A PERSON WHO RAPIDLY CLIMBS A MOUNTAIN FROM SEA LEVEL AND STAYS AT HIGH ALTITUDE (10,000 FEET) FOR SEVERAL WEEKS, UNDERGOES GRADUAL PHYSIOLOGICAL CHANGES THAT PERMIT HIM TO SURVIVE UNDER BETTER CONDITIONS THAN WHEN HE FIRST ARRIVED AT HIS DESTINATION. AFTER ABOUT ONE WEEK THE VENTILATORY RESPONSE INDUCED BY THE LOW pO_2:
A. Is equal as the ventilatory response developed immediately after arriving
B. Is less than the ventilatory response developed immediately after arriving
C. Is greater than the ventilatory response developed immediately after arriving
D. Returns to the level of ventilation present at sea level
Ref. p. 502

SECTION III - PHYSIOLOGY OF THE RESPIRATORY SYSTEM

506. THE CHANGES TAKING PLACE IN THE SUBJECT QUOTED IN THE PREVIOUS QUESTION ARE TO:
   A. Adaptation of the respiratory center to low pO_2
   B. Adaptation of the respiratory center to lower pCO_2
   C. Adaptation of the respiratory center to the lower pH
   D. A change in the rate of basal metabolism
   Ref. p. 502

507. AS A CONSEQUENCE OF SEVERE MUSCULAR EXERCISE ALL OF THE FOLLOWING PHYSIOLOGICAL CHANGES TAKE PLACE IN A NORMAL HUMAN, EXCEPT:
   A. Oxygen consumption increases up to a maximum of about 4 liters/minute
   B. Carbon dioxide production increases up to about 4 liters-5 liters/minute
   C. The mean pCO_2 in the alveolar space does not change significantly from values seen at rest
   D. The mean pO_2 in the alveolar space does not change significantly from values seen at rest
   E. Not more than 90% of blood hemoglobin becomes fully oxygenated in alveolar capillaries due to the rapid passage of blood through the lungs
   Ref. p. 503

508. ACUTE RESPIRATORY DEPRESSION DUE TO THE SO-CALLED "PRESSURE CONUS" FOLLOWS:
   A. Chronic low pressure of the cerebrospinal fluid
   B. Decompression of the spinal canal
   C. Anesthesia
   D. Exposure to high concentrations of carbon dioxide in alveolar air
   Ref. p. 504

509. IN A CASE OF MORPHINE INTOXICATION THE RESPIRATORY DEPRESSION MAY BE OVERCOME WITH CAREFULLY ADMINISTERED DOSES OF:
   A. Halothane
   B. 7% carbon dioxide
   C. Pentobarbital
   D. Picrotoxine
   Ref. p. 504

510. A LOGICAL REASON WHY THE PERIODICITY OF CHEYNE-STOKES BREATHING OCCURS LIES IN THE POSSIBLE ROLE OF A(N):
   A. High cardiac output
   B. Increased rate of oxygen consumption
   C. Slowed circulation from lungs to brain
   D. Permanent damage to neurons of the respiratory center
   Ref. p. 504

511. THE MAIN CAUSE OF INCREASED AIRWAY RESISTANCE DURING EXPIRATION IS:
   A. The change from streamline flow into turbulent flow
   B. An increase in density of the air as it has become fully saturated with water vapor in the alveolar space
   C. A greater rate of flow during expiration
   D. A compression of alveolar bronchioles during expiration
   Ref. p. 509

SECTION III - PHYSIOLOGY OF THE RESPIRATORY SYSTEM 101

512. EMBOLIZATION OF A LARGE BRANCH OF THE PULMONARY ARTERY IS LIKELY TO CAUSE A DISTURBED PULMONARY FUNCTION MAINLY DUE TO:
   A. Excessive ventilation of the increased dead space
   B. Decreased diffusion of gases through the remaining patent alveolar capillary membranes
   C. Increased ventilation/perfusion ratio
   D. Cyanosis   Ref. p. 509

513. BECAUSE OF VARIABLE DEGREES OF ALTERATION OF THE VENTILATION/PERFUSION RATIO IN VARIOUS FORMS OF PULMONARY DISEASES IT IS MOST LIKELY THAT SUCH A RATIO IS LEAST DEVIATED FROM 1.0 IN CASES OF:
   A. Pneumonia
   B. Acute atelectasis
   C. Pulmonary embolism
   D. Bronchial asthma   Ref. p. 511

514. A POLYCYTHEMIC PATIENT IS FOUND TO HAVE 25 GRAMS OF HEMOGLOBIN IN 100 ml OF BLOOD. HIS ARTERIAL BLOOD IS FOUND TO BE 95% SATURATED, BUT HIS FINGER CAPILLARY BLOOD IS 80% SATURATED. HIS CYANOSIS IS EXPLAINED BY THE FACT THAT THE CONCENTRATION OF REDUCED HEMOGLOBIN AMOUNTS TO AT LEAST:
   A. 1 gram/100 ml blood   D. 10 grams/100 ml blood
   B. 3 grams/100 ml blood   E. 15 grams/100 ml blood
   C. 5 grams/100 ml blood   Ref. p. 513

515. THE FOLLOWING STATEMENTS ARE ALL CORRECT, EXCEPT:
   A. Severe dyspnea is never present in a patient, with a respiratory disease, who maintains a normal concentration of oxygen and carbon dioxide in blood
   B. Dyspnea may exist in a patient free of cardiac and respiratory illness
   C. Hypercapnia may lead to dyspnea
   D. Hypoxia may lead to dyspnea
   E. Dyspnea is related to excessive work of muscles involved in the act of breathing   Ref. p. 514

516. MANY CAUSES OF HYPOXIA LEAD TO A CONCOMITANT HYPERCAPNIA. IN MANY CASES, HOWEVER, HYPOXIA MAY EXIST WITHOUT HYPERCAPNIA, WHILE IN OTHERS THE RETENTION OF CARBON DIOXIDE MAY EXIST WITHOUT OXYGEN DEFICIT. HYPERCAPNIA WITHOUT HYPOXIA AS THE UNIQUE SIGN OF RESPIRATORY INSUFFICIENCY SHOULD SUGGEST:
   A. Insufficient ventilation in a person breathing 100% oxygen
   B. An increased carbon dioxide production
   C. A restricted diffusion through the alveolar-capillary membranes
   D. A difficulty in transport due to lack of sufficient hemoglobin
                                        Ref. p. 514

SECTION III - PHYSIOLOGY OF THE RESPIRATORY SYSTEM

517. OXYGEN THERAPY (100% OXYGEN IN A TENT) IS LEAST EFFECTIVE TO CORRECT:
 A. Atmospheric hypoxia
 B. Hypoxia due to hypoventilation
 C. Hypoxia due to impaired diffusion
 D. Hypoxia in carbon monoxide poisoning
 E. Hypoxia due to cyanide poisoning   Ref. p. 514

518. THE OPTICAL SPECTRUM OF OXYHEMOGLOBIN IN ARTERIAL BLOOD WOULD BE LEAST DISTORTED IN CASES OF:
 A. Atmospheric hypoxia
 B. Hypoxia due to hypoventilation
 C. Hypoxia due to impaired diffusion
 D. Hypoxia in carbon monoxide poisoning
 E. Hypoxia due to cyanide poisoning   Ref. p. 514

519. IN THE CASE OF A PATIENT BROUGHT TO THE EMERGENCY WARD OF A HOSPITAL WITH A CLEAR CUT CASE OF MORPHINE POISONING AND HYPOXIA, THE MOST CONTRAINDICATED THERAPEUTIC MEASURE WOULD BE:
 A. To administer 100% oxygen in an oxygen tent
 B. To put the patient in an artificial respirator
 C. To intubate the patient (tracheostomy) and administer oxygen under alternate positive and negative breathing
 D. Give mouth to mouth resuscitation   Ref. p. 515

520. THE DIFFERENT RATE OF REABSORPTION OF INDIVIDUAL GASES CONTAINED IN A BUBBLE OF AIR RETAINED IN A BODY TISSUE EXPLAINS THAT WHEN EQUILIBRIUM IS REACHED THE:
 A. Oxygen pressure equals arterial pO_2
 B. CO_2 pressure equals alveolar pCO_2
 C. Nitrogen pressure is higher than alveolar pN_2
 D. Pressure of water vapor is 10 mms Hg higher than the pH_2O in alveoli   Ref. p. 516

521. AS HIGHER ALTITUDES ARE REACHED IN AN ASCENDING AIRPLANE, THE PILOT IN AN UNPRESSURIZED CABIN IS FACED WITH A(N):
 A. Oxygen concentration in atmospheric air that decreases progressively
 B. Alveolar carbon dioxide concentration which decreases in proportion with the altitude
 C. Oxygen concentration in the atmospheric air which stays as constant as that found at sea level
 D. Pressure of water vapor in the alveolar air which surpasses the pressure of oxygen in alveolar air at 10,000 feet
     Ref. p. 518

522. THE EARLIEST EFFECT OF HYPOXIA IS PROBABLY:
 A. A decreased function of the retinal rods
 B. An increased activity of the carotid chemoreceptors
 C. Drowsiness, lassitude, mental fatigue
 D. Euphoria   Ref. p. 520

SECTION III - PHYSIOLOGY OF THE RESPIRATORY SYSTEM

523. A SUDDEN LOSS OF PRESSURE IN THE AIR IN A SPACE CAPSULE RETURNING TO EARTH'S ATMOSPHERE FROM OUTER SPACE WILL CAUSE IN THEIR OCCUPANTS:
    A. Boiling of the body fluids
    B. Pulmonary fat embolism
    C. Euphoria
    D. An increased fraction of carbamino-hemoglobin at the expense of bicarbonate   Ref. p. 522

524. A PILOT, IN THE SITTING POSITION, PULLING OUT OF A DIVE IN A CONVENTIONAL AIRPLANE, WHO DEVELOPS ANGULAR FORCES EQUAL TO 5 Gs TOWARD HIS SEAT:
    A. Has a cardiac pressure (blood pressure within the left ventricle) lowered to about 40 mm Hg
    B. Has an increased venous return
    C. Cannot faint
    D. Has a high cardiac output   Ref. p. 523

525. THE TIME REQUIRED FOR A PILOT, EXPOSED TO FORCES EQUAL TO 5 Gs, TO EXPERIENCE A BLACKOUT IS:
    A. 5 seconds
    B. 1 minute
    C. 2 minutes
    D. 3 minutes   Ref. Fig. 44-5, p. 524

526. IN A PILOT SUBJECTED TO 3.0 G POSITIVE G FORCES, APPLIED AWAY FROM THE HEART AND TOWARD THE FEET, THE BARORECEPTOR REFLEXES:
    A. Do not allow any recovery because of the large forces involved
    B. Allow a full recovery of arterial pressure and prevent a blackout
    C. Permit a partial recovery with arterial blood pressures maintained at about 50 mms Hg
    D. Cause a paradoxical response with further aggravation of the hypotension   Ref. Fig. 44-5, p. 524

527. THE REASON WHY ASTRONAUTS LIE ON THEIR BACKS FOR TAKE-OFF, MAKING THE DIRECTION OF THE ACCELERATION COINCIDE WITH THE ANTEROPOSTERIOR DIAMETER OF THE CHEST, IS THAT:
    A. Up to 15-20 G forces can be withstood in such a position for many seconds (probably due to least impediment to venous return in both vena cavae, placed perpendicular to the direction of the huge G forces)
    B. Venous return is easier in the position of decubitus for blood returning from the head
    C. The vital capacity is larger in such a position
    D. The baroreceptor reflexes do not have paradoxical responses in such a position   Ref. p. 523

528. THE LEAST EXPECTED RESULT OF THE EXPOSURE OF A PILOT TO -4 G FORCES (OUTSIDE LOOPS) IS:
    A. Increase in the arterial pressure at the level of the heart
    B. An extreme baroreceptor reflex with extreme bradycardia
    C. Rupture of arachnoidal vessels
    D. Ischemia of the retina with blackout
    E. Temporary psychosis due to cerebral edema
    Ref. p. 524

SECTION III - PHYSIOLOGY OF THE RESPIRATORY SYSTEM

529. A PERSON JUMPING WITH A PARACHUTE FROM 5,000 FEET SHOULD EXPECT, THAT WHILE THE PARACHUTE DOES NOT OPEN:
   A. A very fast initial velocity of fall will occur as soon as he leaves the aircraft
   B. An ever increasing velocity as he approaches the earth
   C. That after about 12-15 seconds his velocity of fall will not increase beyond a limiting value
   D. That the resistance imposed by air friction will not modify the acceleration of his fall   Ref. p. 525

530. A PARACHUTE JUMPER WHOSE PARACHUTE HAS PERFORMED PERFECTLY, STRIKES THE EARTH AT A VELOCITY EQUAL TO:
   A. 2 feet/second
   B. 20 feet/second
   C. 100 feet/second
   D. 200 feet/second   Ref. p. 525

531. DUE TO THE SPECIAL FEATURES ASSOCIATED WITH THE RESPONSIVENESS OF THE SEMICIRCULAR CANALS TO ROTATION OF THE BODY:
   A. The aviator is constantly and fully conscious of rotations performed at a uniform velocity
   B. Rotations upward are not perceived at all
   C. A blind aviator is totally incapable of perceiving rotations
   D. An aviator perceives the initial and final phases of uniform rotations, only   Ref. pp. 525-526

532. IN SPACE FLIGHTS, IT SHOULD BE REMEMBERED THAT THE INNER VAN ALLEN RADIATION BELT HAS THE FOLLOWING FEATURES, EXCEPT:
   A. The energy level of particles is as high as 40 Mev
   B. The passage through the inner Van Allen radiation belt could involve the reception of as much as 10 roentgens of radiation
   C. Only a few minutes of orbit in this belt could permit enough radiation to be lethal
   D. It is located between 300-3000 miles away from the surface of the earth
   E. All around the surface of the earth it extends as a covering layer leaving only a passage free of radiation close to the equator   Ref. p. 526

533. IN ADDITION TO THE PRESSURE CAUSED BY THE WEIGHT OF THE AIR IN THE ATMOSPHERE AT SEA LEVEL (760 mm Hg), THE WEIGHT OF WATER AT PROGRESSIVELY DEEPER LEVELS UNDER THE SURFACE OF THE OCEAN (OR FRESH WATER LAKE) INCREASES THE PRESSURE AROUND A SUBMERGED PERSON AT A RATE APPROXIMATELY EQUAL TO:
   A. 1 atmosphere every 13 feet
   B. 1 atmosphere every 23 feet
   C. 1 atmosphere every 33 feet
   D. 1 atmosphere every 53 feet
   E. 1 atmosphere every 93 feet   Ref. p. 529

SECTION III - PHYSIOLOGY OF THE RESPIRATORY SYSTEM

534. THE RECIPROCAL RELATIONSHIP BETWEEN PRESSURE OF A CERTAIN GIVEN QUANTITY OF A GAS AND THE VOLUME IT OCCUPIES (BOYLE'S LAW) PERMITS US TO ANTICIPATE THAT 300 FEET UNDER THE SURFACE OF THE OCEAN (10 ATMOSPHERES) A LUNG CAPACITY EQUAL TO 8 LITERS AT SEA LEVEL WILL BE DECREASED TO:
    A. 4 liters
    B. 6 liters
    C. 2 liters
    D. 0.8 liters
    E. 0.08 liters
    Ref. p. 529

535. NITROGEN, WHICH COMPRISES 78% OF THE MOLECULES OF ATMOSPHERIC AIR AND WHICH IS LARGELY AN INERT GAS AT SEA LEVEL:
    A. Does not cause any damage to the human body at any depth
    B. Causes narcosis at about 33 feet under the surface of the sea when its pressure equivalent, in a person breathing air, equals 1186 mm Hg (2 atmospheres x 760 mms Hg/atm x 0.78 FN_2=1186)
    C. Causes very rapid changes at high pressure due to its rapid solubility in tissue fluids
    D. Will cause narcosis 1 hour after exposure at 300 feet (10 atmospheres)
    Ref. p. 530

536. BREATHING PURE OXYGEN FROM A TANK AT A DEPTH EQUAL TO 66 FEET (3 ATMOSPHERES) BELOW THE SURFACE OF THE OCEAN WILL RESULT IN:
    A. No physiological change for at least 24 hours
    B. Recovery from nitrogen narcosis at such depth
    C. Convulsions if the exposure lasts more than an hour
    D. Euphoria
    Ref. p. 530

537. A PATIENT CONFINED TO BED, BREATHING 100% OXYGEN IN AN OXYGEN TENT, AT SEA LEVEL, WILL DEVELOP:
    A. No damage whatever
    B. Pulmonary edema after 24-48 hours
    C. Cerebral vasodilatation, after a few minutes
    D. Narcosis after 24-48 hours
    E. Euphoria after 24-48 hours
    Ref. p. 531

538. ALL THE FOLLOWING STATEMENTS CONCERNING HELIUM ARE CORRECT, EXCEPT:
    A. Helium is less narcotic than nitrogen
    B. Helium has a lower density than nitrogen
    C. Helium reduces the resistance to air flow
    D. Helium is less soluble than nitrogen in tissues
    E. Helium diffuses less rapidly through tissues than nitrogen
    Ref. p. 531

539. IN PREPARING A GAS MIXTURE INTENDED FOR BREATHING AT HIGH PRESSURES UNDER THE LEVEL OF THE SEA, IT IS ADVANTAGEOUS TO REPLACE NITROGEN BY:
    A. Oxygen
    B. Carbon dioxide
    C. Argon
    D. Helium
    E. Hydrogen
    Ref. p. 531

SECTION III - PHYSIOLOGY OF THE RESPIRATORY SYSTEM

540. A VERY OBESE PERSON EXPOSED FOR SEVERAL HOURS TO 4 ATMOSPHERES OF AIR PRESSURE:
   A. Should be decompressed more slowly than a lean person
   B. Has a greater fraction of nitrogen dissolved in his body water than in his fat, due to a partition coefficient water/adipose tissue for nitrogen greater than one
   C. Will have 10 liters of nitrogen (sea level volume of nitrogen) dissolved in the whole body (as compared to 1 liter at sea level)
   D. Will be better protected from the "bends" than a lean person
   Ref. p. 531

541. NITROGEN IS PREFERABLE OVER HELIUM AS THE GAS TO BE MIXED WITH OXYGEN IN MIXTURES USED IN UNDERWATER WORK WHEN:
   A. The person that will use the mixture is obese
   B. The depth is close to 300 feet under the surface
   C. The length of the stay underwater is about 2 hours
   D. Decompression is to be made very slowly
   E. The depth is shallow and the duration of stay is a short one
   Ref. p. 534

542. ASSUME THAT A PERSON WORKING 300 FEET UNDER THE SURFACE OF THE OCEAN HAS A TIDAL VOLUME EQUAL TO 0.2 LITERS. THE CARBON DIOXIDE EXHALED WITH EACH TIDAL MOVEMENT WILL BE EFFICIENTLY LOST TO THE SURFACE, IF AT THE SURFACE A REPLACEMENT VOLUME PUSHED UNDER PRESSURE TOWARD THE MAN'S BREATHING HELMET EQUALS AT LEAST:
   A. 0.2 liters per respiration
   B. 0.5 liters per respiration
   C. 1.0 liters per respiration
   D. 2.0 liters per respiration
   E. 4.0 liters per respiration
   Ref. p. 534

543. THE MAIN REASON WHY THE MAXIMAL BREATHING CAPACITY DECREASES SIGNIFICANTLY IN A PERSON WORKING UNDER THE SURFACE OF WATER IS THAT:
   A. The respiratory muscles work less efficiently when subjected to external pressure higher than atmospheric
   B. The hemoglobin is less well saturated
   C. Bronchi are partially collapsed at high pressure
   D. The air has a greater density in proportion to the depth
   Ref. p. 534

544. A PERSON WHO ESCAPES FROM A SUBMARINE SWIMMING TO THE SURFACE WITH A SELF-CONTAINED UNDERWATER BREATHING APPARATUS SHOULD BE CAREFUL TO:
   A. Inspire deeply to increase his alveolar oxygen concentration
   B. Exhale continuously
   C. Ascend as rapidly as possible
   D. Move his extremities very actively to increase the release of dissolved nitrogen
   Ref. p. 536

SECTION III - PHYSIOLOGY OF THE RESPIRATORY SYSTEM

545. THE MECHANISM OF DEATH IS DIFFERENT IN CASES OF PERSONS DROWNED IN FRESH WATER AS OPPOSED TO PERSONS WHO DROWN AT SEA. IT SEEMS THAT DEATH IN FRESH WATER:
   A. Is associated with a rapid reduction of the extracellular fluid volume due to rapid osmotic passage of water into cells
   B. Results in acute hyperkalemia due to massive hemolysis
   C. Is slower (15 minutes) than at sea (1 minute)
   D. Is associated with a very high resting membrane potential of the cardiac muscles
   Ref. pp. 536-537

SECTION IV - PHYSIOLOGY OF THE NERVOUS SYSTEM

IN THE FOLLOWING QUESTIONS (546-564) A STATEMENT IS FOLLOWED BY FOUR POSSIBLE ANSWERS. ANSWER BY USING THE KEY OUTLINED BELOW:
A. If only 1, 2 and 3 are correct
B. If only 1 and 3 are correct
C. If only 2 and 4 are correct
D. If only 4 is correct
E. If all are correct

546. MOLECULES OF THE TRANSMITTER SUBSTANCE STORED IN SYNAPTIC VESICLES ARE:
1. Probably synthesized in situ with the intervention of ATP
2. Partially released when the presynaptic membrane is depolarized
3. Released in proportion to ionic calcium concentration in the extracellular fluid
4. Not released at a normally rapid rate if there is an excess of magnesium ions in the extracellular fluid
Ref. p. 67

547. SUBSTANCES WHICH PROBABLY HAVE A ROLE AS TRANSMITTERS IN THE CENTRAL AND PERIPHERAL NERVOUS SYSTEM INCLUDE:
1. Acetylcholine
2. Serotonin
3. Norepinephrine
4. L-glutamic acid
Ref. p. 68

548. A PRE-SYNAPTIC DEPOLARIZATION INDUCES IN THE POST-SYNAPTIC MEMBRANE OF NEURONS A(N):
1. Increased permeability to both sodium and potassium ions
2. Diffusion of sodium ions of sufficient magnitude to reach the value of sodium equilibrium potential, under all conditions of excitation of the post-synaptic membrane
3. Excitatory post-synaptic potential (EPSP) which lasts much longer (about 15 milliseconds) than an action potential
4. Much greater diffusion of potassium ions than of sodium ions
Ref. p. 68

549. THE THRESHOLD OF EXCITATION OF A NEURON:
1. Is reached when the excitatory post-synaptic potential (EPSP) reaches a mean value of about 1-2 millivolts less negative than the resting potential
2. Can be reached by either temporal or spatial summation
3. Is higher in the initial segment of the axon than in the rest of the soma of the neuron
4. Varies for different neurons, between 5-25 mV on the positive side of the resting potential
Ref. p. 70

SECTION IV - PHYSIOLOGY OF THE NERVOUS SYSTEM

550. IF THE POST-SYNAPTIC NEURON IS DEPOLARIZED BY PRE-SYNAPTIC IMPULSES THE:
   1. Rate of discharge of the post-synaptic neuron is proportional to the magnitude of depolarization of the EPSP
   2. Rate of discharge, for a given EPSP, will vary from one neuron to another
   3. Frequency of discharge may rise from zero (at rest) to 500 and to even 1000 action potentials per second
   4. Peak of the action potential is the same regardless of the magnitude of the depolarization caused by the (summated) post-synaptic potentials
   Ref. p. 70

551. INHIBITORY POST-SYNAPTIC POTENTIALS (IPSP):
   1. Are caused by either stimulatory or inhibitory synapses which can give rise to both excitation and inhibition
   2. May occur because of a selective increased permeability to potassium ions
   3. Are associated with a large flux of chloride ions
   4. Last 10-15 milliseconds   Ref. p. 71

552. IN A GIVEN NEURON IN THE CENTRAL NERVOUS SYSTEM:
   1. Inhibitory and excitatory synapses may occur adjacent to one another
   2. The inhibitory transmitter may relate to gamma-amino-butyric acid (GABA)
   3. An incoming IPSP is usually originated in a specific inhibitory neuron
   4. An incoming EPSP is usually originated in a specific excitatory neuron
   Ref. p. 72

553. CONDUCTION OF NERVE IMPULSES FROM ONE NEURON TO THE NEXT:
   1. Occurs through synapses only in one direction and never in the opposite one
   2. Takes place with a synaptic delay equal to about 5 milliseconds
   3. May slow down due to "fatigue" in the synaptic transmission
   4. Is not inhibited by pre-synaptic inhibition
   Ref. pp. 72-73

554. IN THE PHENOMENON KNOWN AS POST-TETANIC FACILITATION:
   1. Rapid pre-synaptic stimulation facilitates a tetanic response in the post-synaptic cell
   2. There is facilitation of conduction in the presynaptic neuron after a tetanus
   3. The synaptic delay is reduced to half its control values after a tetanus
   4. Repetitive stimuli to a pre-synaptic neuron for a short time will cause the post-synaptic neuron to become more responsive to pre-synaptic influence   Ref. p. 73

555. SYNAPTIC TRANSMISSION IS:
   1. Facilitated by alkalosis
   2. Inhibited by hypoxia
   3. Facilitated by coffee or tea
   4. Facilitated by strychnine (which acts by increasing the excitability of neurons very markedly)   Ref. p. 73

SECTION IV - PHYSIOLOGY OF THE NERVOUS SYSTEM

556. A PROGRESSIVE INCREASE IN THE INTENSITY OF STIMULI APPLIED TO A GIVEN AREA OF THE SKIN:
    1. Causes a sensation of greater intensity because more stimuli are transmitted in the unit time by a given axon
    2. Causes a sensation of greater intensity because a larger number of nerve fibers participate in the conduction of the stimulus
    3. Causes a continuous and smooth change in post-synaptic responses (neuronal or muscular)
    4. Cannot be spatially oriented in the ascending nerve tracts
    Ref. pp. 547-548

557. IN A NEURONAL POOL:
    1. A subthreshold stimulus applied on several pre-synaptic terminals may converge and summate to depolarize a neuron sufficiently to reach threshold
    2. The area where all neurons discharge is the liminal zone
    3. The area where neurons are facilitated but do not discharge is the subliminal zone
    4. A given neuron whose discharge is facilitated by some afferent signals may be also inhibited by other afferents
    Ref. pp. 549-550

558. MOTOR NEURONS IN THE SPINAL CORD MAY BE THE FOCUS OF CONVERGENCE OF STIMULI ORIGINATED IN:
    1. The cerebral cortex
    2. Internuncial cells in the spinal cord
    3. Cells in the dorsal root ganglion
    4. The reticular formation
    Ref. p. 551

559. THE PHENOMENON OF DIVERGENCE OF CONDUCTION OF NEURAL STIMULI:
    1. Does not exist in the central nervous system and is restricted to the autonomic nervous system
    2. Weakens the effector's response
    3. May allow a limit of 10-12 neurons to participate
    4. May be exemplified by the proprioceptive pathways into the cerebellum and into the thalamus
    Ref. p. 551

560. A SINGLE BRIEF STIMULUS, LASTING LESS THAN 1 MILLISECOND, APPLIED TO AN AFFERENT FIBER, MAY RESULT IN A PROLONGED OUTPUT SIGNAL FROM A NEURON THAT FORMS PART OF A NEURONAL POOL DUE TO ONE OR SEVERAL OF THE FOLLOWING MECHANISMS:
    1. The fact that the post-synaptic potentials actually last many milliseconds (up to 15 milliseconds and even more)
    2. After-discharge occurring in parallel circuits which converge on one output neuron
    3. After-discharge occurring in reverberating circuits
    4. Adaptation of the output neuron
    Ref. pp. 553-554

561. RECEPTORS WHICH CAN RESPOND TO MECHANICAL DEFORMATIONS INCLUDE:
    1. Free nerve endings
    2. Pacinian corpuscles
    3. Ruffini's endings
    4. Carotid baroreceptors
    Ref. p. 558

SECTION IV - PHYSIOLOGY OF THE NERVOUS SYSTEM 111

562. RECEPTORS WHICH CAN RESPOND TO CHANGES IN THE CHEMICAL COMPOSITION OF THE PLASMA WATER PERFUSING THEM MAY BE FOUND IN THE:
1. Hypothalamic neurons in the lateral area
2. Carotid bodies
3. Supraoptic nuclei
4. Olfactory cells
Ref. p. 558

563. SENSATIONS OF A VERY DIFFERENT NATURE (PAIN, COLD, SIGHT) ARE ALL TRANSMITTED BY ONLY ONE TYPE OF PHYSICAL CHANGE THAT TAKES PLACE IN THE MEMBRANE OF NEURONS AND AXONS (DEPOLARIZATION). THE REASON FOR THIS LIES IN THE FACT THAT THE:
1. Type of sensation depends on the intensity of the stimulation
2. Modality of sensation depends on the degree of after-discharge, stimulation and inhibition of the various neuronal pools that form part of the sensory tract
3. Modality of sensation depends on the type of neurotransmitter present in the synapses involved
4. Type of sensation depends on the area of the central nervous system where the pathways end
Ref. p. 558

564. THE EARLIEST PHYSICAL CHANGES IN THE PERIPHERAL NERVOUS SYSTEM THAT EVENTUALLY RESULTS IN A SENSATION OCCUR THROUGH:
1. A probable increase in ionic permeability of deformed receptors
2. The appearance of a local electrical current
3. The spreading of such current centrally
4. An adaptation process of variable duration
Ref. pp. 559-560

FOR EACH OF THE FOLLOWING MULTIPLE CHOICE QUESTIONS, SELECT THE ONE MOST APPROPRIATE ANSWER:

565. THE RELATIONSHIP BETWEEN THE MAGNITUDE OF THE GENERATOR POTENTIAL FOR THE MUSCLE SPINDLE AND THE FREQUENCY OF RESPONSE IS A(N):
A. Reciprocal one
B. Directly proportional one
C. Logarithmic one
D. Exponential one
Ref. Fig. 48-4, p. 560

566. THE AMPLITUDE OF GENERATOR POTENTIALS OBSERVED IN THE PACINIAN CORPUSCLES DEPENDS ON THE STIMULUS STRENGTH:
A. At very high levels of stimulus strength there is an ever increasing amplitude of the potential
B. At very low levels of stimulus strength the amplitude is equally as high as at high levels of stimulus strength
C. The amplitude of generator potentials increases rapidly at first, reaching a maximal value after a certain value of stimulus strength has been reached
D. The amplitude can reach a maximum of 1 millivolt
Ref. p. 559

SECTION IV - PHYSIOLOGY OF THE NERVOUS SYSTEM

567. THE PHENOMENON OF ADAPTATION IS MAXIMAL IN:
    A. Pacinian corpuscles
    B. Hair base receptors
    C. Muscle spindles
    D. Joint capsule receptors   Ref. Fig. 48-5, p. 561

568. FAST ADAPTING RECEPTORS INCLUDE THOSE FOUND IN THE:
    A. Macula in the vestibular apparatus   D. Carotid baroreceptors
    B. Sound receptors of the ear   E. Tactile receptors
    C. Pain receptors   Ref. p. 561

569. THE CAPACITY TO DISCRIMINATE AN INCREMENT IN INTENSITY OF STIMULUS IS GREATER:
    A. In the higher range of stimulation
    B. In the middle range of stimulation
    C. In the lower range of stimulation
    D. Equal in all sections of the range of intensities
                                          Ref. p. 563

570. IN THE ABILITY TO RECOGNIZE CHANGES IN STIMULI, THE RATIO $\frac{\text{CHANGE IN STIMULUS STRENGTH}}{\text{MAGNITUDE OF INITIAL STIMULUS STRENGTH}}$ IS:
    A. Essentially a constant one
    B. Increasingly larger
    C. Increasingly smaller
    D. An exponential function of the magnitude of the initial stimulus strength   Ref. pp. 562-563

571. THE SENSATIONS OF PROPRIOCEPTION ARE CARRIED TO THE CENTRAL NERVOUS SYSTEM BY FIBERS OF:
    A. Type A (alpha)
    B. Type A (delta)
    C. Type B
    D. Type C   Ref. Table 48-2, pp. 563-564

572. SENSATIONS OF PAIN, HEAT, COLD ARE CARRIED TO THE CENTRAL NERVOUS SYSTEM BY FIBERS OF:
    A. Type A (alpha)
    B. Type A (delta)
    C. Type B
    D. Type C   Ref. Table 48-2, pp. 563-564

573. PREGANGLIONIC AUTONOMIC FIBERS ARE OF:
    A. Type A (alpha)
    B. Type A (delta)
    C. Type B
    D. Type C   Ref. Table 48-2, pp. 563-564

574. POST-GANGLIONIC AUTONOMIC FIBERS ARE OF:
    A. Type A (alpha)
    B. Type A (delta)
    C. Type B
    D. Type C   Ref. Table 48-2, pp. 563-564

SECTION IV - PHYSIOLOGY OF THE NERVOUS SYSTEM

575. SENSATIONS OF PROPRIOCEPTION (CARRIED AT MEAN VELOCITY OF 100 METERS/SEC) MAY BE CARRIED FROM FEET TO BRAIN STEM (1.5 METERS) IN:
    A. 0.015 second
    B. 0.15 second
    C. 1.5 second
    D. 15 seconds   Ref. p. 563

576. SENSATIONS OF PAIN IN THE FEET, CARRIED IN FIBERS OF TYPE C (VELOCITY EQUAL TO 0.5 METERS/SECOND) WILL REACH THE BRAIN STEM (1.5 METERS FROM FEET TO HEAD) IN:
    A. 0.03 seconds
    B. 0.3 seconds
    C. 3.0 seconds
    D. 30 seconds   Ref. p. 563

577. THE MOST RAPIDLY ADAPTING OF THE TACTILE RECEPTORS IS KNOWN AS:
    A. Meissner's corpuscle   D. Ruffini end organ
    B. Merkel's disc   E. Pacinian corpuscle
    C. Hair end organ   Ref. p. 566

578. SPECIALIZED SENSORY RECEPTORS (PACINIAN, MEISSNER, ETC.) TRANSMIT THEIR ELECTRICAL CHANGES TO THE CENTRAL NERVOUS SYSTEM THROUGH FIBERS OF THE CLASS:
    A. A (beta), B, C
    B. B, C
    C. A (beta), C
    D. A (beta), only   Ref. p. 566

579. VELOCITY OF CONDUCTION FOR SIGNALS ORIGINATED IN THE ABOVE MENTIONED SPECIALIZED SENSORY RECEPTORS IS ON THE ORDER OF:
    A. 30-50 meters/second
    B. 3-14 meters/second
    C. 1-2 meters/second
    D. 120 meters/second   Ref. p. 564

580. ACTION POTENTIALS ORIGINATED IN FREE NERVE ENDINGS TRAVEL THROUGH NERVES ENDOWED WITH A VELOCITY OF ABOUT:
    A. 100-120 meters/second
    B. 40-50 meters/second
    C. 20-30 meters/second
    D. 1-8 meters/second   Ref. p. 566

581. THE MOST RAPID VIBRATORY STIMULI TRANSMIT THEIR SIGNALS VIA:
    A. Pacinian corpuscles
    B. Meissner corpuscles
    C. Ruffini endings
    D. Expanded tip receptors   Ref. p. 566

SECTION IV - PHYSIOLOGY OF THE NERVOUS SYSTEM

582. THE PROGRESSIVE ANGULATION OF A JOINT ELICITS:
    A. A non-adapting stream of impulses from specialized receptors (Pacini)
    B. Adapting impulses that appear at different angles and then disappear at a further angulation
    C. Progressive increase of firing of impulses as the angle decreases
    D. Progressive decrease of firing of impulses as the angle decreases
    Ref. Fig. 49-1, p. 567

583. THE MOST RAPIDLY TRAVELING SOMATOSENSORY INFORMATION TO THE BRAIN TRAVELS WITH THE:
    A. Dorsal columns
    B. Ventral spino-thalamic tracts
    C. Dorsal spino-thalamic tracts
    D. Sympathetic afferents   Ref. p. 568

584. SOMATOSENSORY INFORMATION REACHES THE BRAIN THROUGH:
    A. Dorsal columns of the cord and spinothalamic tracts
    B. Dorsal columns alone
    C. Ventral spinothalamic tract alone
    D. Dorsal spinothalamic tract alone   Ref. pp. 567-568

585. THE DORSAL COLUMN AFFERENT SYSTEM TRANSMITS INFORMATION CONCERNING:
    A. Fine touch, vibration, kinesthetic sensations
    B. Pain, sexual sensations
    C. Thermal sensations
    D. Crude touch, tickle   Ref. p. 568

586. THE SPINOTHALAMIC AFFERENT SYSTEM TRANSMITS INFORMATION CONCERNING:
    A. Pain and temperature
    B. Fine touch
    C. Vibrations
    D. Kinesthetic sensations   Ref. p. 568

587. DORSAL COLUMN FIBERS ASCEND TO SYNAPSE AT THE:
    A. Thalamus
    B. Trigeminal nuclei
    C. Cuneate and gracilis nuclei
    D. Somatosensory cortex   Ref. p. 569

588. THE TRANSECTION OF THE MEDIAL LEMNISCI ON THE RIGHT SIDE OF THE BRAIN STEM WILL SELECTIVELY ABOLISH SENSATIONS OF:
    A. Pain in the opposite side of the body
    B. Pain in the same side of the body
    C. Fine touch on the opposite side of the body
    D. Sexual sensations   Ref. p. 569

589. THE MAIN FINAL DESTINATION OF ELECTRICAL IMPULSES THAT CARRY SENSATIONS OF FINE TOUCH IS THE:
    A. Cuneate and gracilis nuclei
    B. Ventrobasal nuclei of the thalamus
    C. Post-central gyrus of the cerebral cortex
    D. Cerebellar cortex   Ref. p. 569

SECTION IV - PHYSIOLOGY OF THE NERVOUS SYSTEM

590. SOMATOTOPIC DISTRIBUTION OF THE ORIENTATION OF FIBERS CARRYING SENSATIONS OF FINE TOUCH IS ARRANGED SO THAT THE:
    A. Fibers from the lower part of the body lie in the lateral portion of the dorsal column, in the spinal cord
    B. Caudal part of the body coincides with the medial part of the ventrobasal nuclei of the thalamus
    C. Lower extremity is projected to the medial portion of the somesthetic cortex (sensory area I)
    D. Smallest portion of the somatosensory cortex corresponds to the projection of the lips of the mouth   Ref. pp. 569-570

591. THE SOMATIC SENSORY AREA II IS LOCATED:
    A. Slightly behind and below the lower end of the post-central gyrus
    B. In the upper part of the post-central gyrus
    C. In the medial portion of the post-central gyrus
    D. Near the medial and upper end of the motor cortex
    Ref. p. 570

592. A PERSON WHO HAS HAD A DESTRUCTION OF THE SOMATIC SENSORY AREA I ON THE RIGHT SIDE WILL SHOW:
    A. Inability to judge shapes of objects on the left side
    B. Total lack of temperature sensations on the same side
    C. Total lack of pain sensations on the same side
    D. Inability to judge shapes of objects on the right side
    Ref. p. 571

593. CORTICAL NEURONS IN THE SOMESTHETIC CORTEX RESPOND TO SENSORY STIMULI IN SUCH A WAY THAT:
    A. When a point in the skin is touched there is one point in the somesthetic cortex which is more intensively stimulated than neighboring areas
    B. Adjacent areas to the maximally stimulated area in the cortex are never inhibited
    C. Points with maximal responses vary from time to time according to the magnitude of the force applied
    D. No specific pattern of response exists either to placement of stimuli or intensity of stimulation   Ref. pp. 571-572

594. THE SENSATION INDUCED BY PLACING A VIBRATING TUNING FORK (512 CYCLES/SEC) OVER THE SKIN THAT COVERS THE TIBIA IS TRANSMITTED THROUGH THE:
    A. Dorsal column system
    B. Ventral spino-thalamic system
    C. Spino-cerebellar system
    D. Dorsal spino-thalamic system   Ref. p. 572

SECTION IV - PHYSIOLOGY OF THE NERVOUS SYSTEM

595. SENSATIONS INDUCED BY PROGRESSIVELY ANGULATING A JOINT ARE MEDIATED BY THALAMIC NEURONS IN SUCH A WAY THAT:
    A. The same group of neurons fire at an increasing rate at low and wide angles, with a lowest value at a specific angle
    B. A given group of neurons fire more rapidly as the joint angle is increased while another group of neurons fire more rapidly as the joint angle is decreased
    C. There is no change in rate of firing of neurons when the joint angle varies in any of the studied groups of neurons in the thalamus
    D. Only the rate of change of angulation of the joint is capable of inducing a change in rate of firing of thalamic neurons
    Ref. Fig. 49-8, p. 573

596. THE FIRST SYNAPSES OF FIBERS CONDUCTING PAIN AND TEMPERATURE TAKES PLACE AT THE LEVEL OF THE:
    A. Gray matter of the dorsal horns of the spinal cord
    B. Anterior gray columns of the spinal cord
    C. Nucleus cuneatus
    D. Thalamus
    Ref. p. 573

597. THE FIBERS COMPOSING THE LATERAL SPINO-THALAMIC TRACT:
    A. Are decussated in the posterior commissure of the gray matter of the spinal cord
    B. End in the anterior thalamic nucleus
    C. Terminate in the intralaminar nuclei of the thalamus, separately from the ventral spinothalamic tract
    D. Do not send collaterals to the reticular formation of the brain stem
    Ref. p. 574

598. SEXUAL SENSATIONS ASCEND TO THE BRAIN THROUGH THE:
    A. Dorsal column system
    B. Spino-cerebellar tract
    C. Spinothalamic tracts
    D. Vestibulo-spinal tracts
    Ref. p. 575

599. SYNAPTIC ACTIVITY OF NEURONS INVOLVED IN TRANSMISSION OF AFFERENT SENSORY INPUT:
    A. Is influenced by efferent signals originated in the sensory cortex
    B. Is modified by motor efferents
    C. Depends exclusively on the magnitude of stimuli and of metabolic conditions of the synapsing neurons, without interference from other regions of the CNS
    D. Is influenced by the vestibulo-spinal tract
    Ref. p. 575

600. DERMATOMES ARE ARRANGED IN SUCH A WAY THAT:
    A. Transection of the dorsal root that corresponds to spinal cord segment Lumbar 5, will cause anesthesia of the posterior portion of the thigh
    B. A lesion localized in dorsal root ganglia Cervical 4 and Cervical 5 will cause anesthesia of the hand
    C. A surgical transection of roots Thoracic 9 and Thoracic 10 will cause anesthesia of the anterior portion of the thighs
    D. A disc compressing spinal cord segment Sacral 1, on the right, will cause sensory changes in the sole of the right foot
    Ref. Fig. 49-10, p. 575

SECTION IV - PHYSIOLOGY OF THE NERVOUS SYSTEM 117

601. HEATING THE SKIN SURFACE MAY CAUSE PAIN, ONCE THE "THRESHOLD" FOR THE SENSATION OF PAIN IS REACHED. THIS THRESHOLD:
    A. Is reached in some humans only after the skin has been heated to 60^0 C
    B. Is quantitatively higher in culturally less advanced human groups
    C. Has a range that extends from 43.5^0 C - 47^0 C
    D. Equals the intensity of stimulation that causes pain after one second of applying the stimulus   Ref. Fig. 50-2, p. 578

602. ALL CASES OF PAIN ARE BELIEVED TO BE PRODUCED BY CELLULAR DAMAGE, WHICH APPARENTLY RELEASES:
    A. Bradykinin
    B. An excess of norepinephrine
    C. An excess of acetylcholine
    D. Heparin
    E. Adenosine-tri-phosphate (ATP)   Ref. p. 579

603. COLLATERALS OF THE SPINO-THALAMIC TRACTS TO THE RETICULAR FORMATION OF THE BRAIN STEM ARE BELIEVED TO:
    A. Have no effect on the quality or intensity of pain sensations
    B. Increase the level of excitability of the brain
    C. Cause scotomata
    D. Decrease reflex activity   Ref. p. 581

604. A CORDOTOMY INTENDED TO RELIEVE INTRACTABLE PAIN ORIGINATING IN THE LEFT LEG, SHOULD CONSIST OF TRANSECTION OF THE:
    A. Right thoracic anterolateral quadrant of the spinal cord
    B. Right dorsal columns (fasciuli gracilis and cunneatus)
    C. Left thoracic anterolateral quadrant of the spinal cord
    D. Left dorsal columns (fasciculi gracilis and cunneatus)
    E. Whole spinal cord at the level of segment Lumbar 3
        Ref. p. 581

605. THE SOMESTHETIC CORTEX:
    A. Is indispensable for the conscious sensation of pain
    B. Will never give rise to pain, if electrically stimulated
    C. If removed, will still leave an animal capable of feeling pain
    D. Especially the sensory area II, has little to do with sensations of pain
        Ref. p. 581

606. THE ALMOST IMMEDIATE SENSATION OF PAIN THAT FOLLOWS A PAINFUL TRAUMA SEEMS TO BE CARRIED TO THE CENTRAL NERVOUS SYSTEM BY A SET OF FIBERS DIFFERENT FROM THOSE THAT CARRY THE MORE DELAYED UNPLEASANT SENSATION, ALSO DESCRIBED AS "PAINFUL." IT IS BELIEVED THAT THE "FAST" SHARP SENSATIONS OF PAIN ARE CARRIED BY FIBERS THAT BELONG TO:
    A. Type A (delta) fibers
    B. Type B fibers
    C. Type C fibers
    D. Type A (alpha) fibers   Ref. p. 580

SECTION IV - PHYSIOLOGY OF THE NERVOUS SYSTEM

607. IN THE BRAIN STEM TWO PATHWAYS ARE ANATOMICALLY RELATED TO THE RETICULAR FORMATION AND SEEM TO MODIFY THE PAIN CONDUCTION THROUGH THE ADJACENT SPINOTHALAMIC TRACTS. THE ONE BELIEVED TO BE INVOLVED IN THE SUPPRESSION OF PAIN IS THE:
    A. Central gray pathway
    B. Central tegmental pathway
    C. Olivocerebellar tract
    D. Corticobulbar tract   Ref. p. 581

608. REFERRED PAIN FROM THE HEART TO THE SKIN OF THE PRE-CORDIUM PROBABLY TAKES PLACE BECAUSE OF AN OVERLAP OF VISCERAL AND CUTANEOUS SYNAPSES AT THE LEVEL OF THE:
    A. Somesthetic cortex
    B. Thalamus
    C. Nucleus gracilis and cuneatus
    D. Spinal cord   Ref. p. 582

609. ALL OF THE FOLLOWING MAY GIVE RISE TO VISCERAL PAIN, EXCEPT:
    A. Ischemia
    B. Distention of a hollow viscus
    C. Spasm of a hollow viscus
    D. Incision with a sharp blade   Ref. pp. 582-583

610. A SURGEON MAKING SMALL INCISIONS IN VARIOUS AREAS OF THE ABDOMEN IN A PERSON ANESTHETIZED WITH LOCAL ANESTHESIA IS LIKELY TO ELICIT SHARP PAIN IN UNANESTHETIZED PORTIONS OF THE:
    A. Small intestine
    B. Large intestine
    C. Parietal peritoneum
    D. Liver parenchyma   Ref. p. 583

611. VISCERAL SENSATIONS REACH THE SPINAL CORD VIA:
    A. Efferent sympathetic fibers
    B. Somatic sensory nerves that traverse through the chest or abdominal wall
    C. Motor nerves to muscles of the chest or abdominal wall
    D. Afferent fibers of the autonomic nervous system
                                          Ref. p. 583

612. PAIN ORIGINATED IN THE PELVIC CAVITY (UTERUS, BLADDER, RECTUM) REACHES THE SPINAL CORD THROUGH:
    A. Sacral parasympathetic fibers
    B. Vagal fibers that reach the pelvis
    C. Phrenic nerves
    D. Collaterals from the sciatic nerves   Ref. p. 583

SECTION IV - PHYSIOLOGY OF THE NERVOUS SYSTEM   119

613. THE INFLAMMATION OF THE APPENDIX STARTS USUALLY IN THE WALL OF THE APPENDIX CLOSE TO THE LUMEN AND EXTENDS CENTRIFUGALLY, INVOLVING EVENTUALLY THE PARIETAL PERITONEUM THAT COVERS THE CECUM. BECAUSE OF DIFFERENT AFFERENT PATHWAYS FROM THE MESOAPPENDIX AND FROM THE PARIETAL PERITONEUM THAT COVERS THE CECUM, THE INITIAL PAIN OF AN APPENDICITIS IS LIKELY TO BE FELT AT THE LEVEL OF THE:
   A. Right upper quadrant
   B. Left lower quadrant
   C. Right lower quadrant
   D. Umbilical region
   E. Right posterior lumbar region   Ref. Fig. 50-8, p. 584

614. THE STELLATE GANGLION IS IN THE PATH OF AFFERENT FIBERS THAT MAY CONDUCT PAINFUL STIMULI ORIGINATING IN THE:
   A. Stomach   D. Heart
   B. Duodenum   E. Diaphragm
   C. Gallbladder   Ref. p. 584

615. GALLBLADDER PAIN IS NOT INFREQUENTLY REFERRED TO THE:
   A. Left upper quadrant
   B. Tip of the left scapula
   C. Tip of the right scapula
   D. Left shoulder   Ref. p. 585

616. BECAUSE OF THE RELATIONS OF THE PANCREAS TO THE PERITONEUM AN ACUTE NECROSIS OF THE PANCREAS IS LIKELY TO BE FELT:
   A. Diffusely over the right upper quadrant
   B. Diffusely over the left upper quadrant
   C. In the precordium
   D. Directly behind the pancreas in the back
                                 Ref. p. 585

617. PAIN ORIGINATED IN LESIONS OF THE RIGHT KIDNEY ARE SOMETIMES RADIATED TO THE:
   A. Left shoulder
   B. Right shoulder
   C. Left groin
   D. Right testicle   Ref. p. 585

618. STRICT UNI- OR BI-DERMATOMAL DISTRIBUTION OF PAIN SHOULD SUGGEST THE DIAGNOSIS OF:
   A. Leprosy
   B. Syphilis
   C. Tuberculosis
   D. Herpes zoster
   E. Lead induced neuritis   Ref. p. 586

SECTION IV - PHYSIOLOGY OF THE NERVOUS SYSTEM

619. FOLLOWING A BULLET WOUND THAT DAMAGES THE SPINAL CORD, THE SUPPRESSION OF MOTOR FUNCTION AND TOUCH SENSATION ON THE RIGHT LOWER EXTREMITY, PLUS THE LOSS OF PAIN AND TEMPERATURE ON THE LEFT SIDE SUGGESTS A DIAGNOSIS OF:
  A. Total transection of the spinal cord at the level of lumbar segment I
  B. Hemorrhage of the anterior horns of the gray matter
  C. Compression by a clot over the anterior roots of lumbar segment I
  D. Brown-Sequard syndrome
  E. Traumatic polyneuritis   Ref. p. 586

620. OCCLUSION OF THE ARTERY THAT PERFUSES THE RIGHT POSTEROVENTRAL PORTION OF THE THALAMUS IS LIKELY TO CAUSE, AFTER SEVERAL MONTHS:
  A. Permanent suppression of all sensations on the left side
  B. Permanent suppression of all sensations on the right side
  C. Occassional lancinating pains, elicited by mild stimulations such as touch
  D. Anosmia   Ref. p. 587

621. EXQUISITE AND INTENSE PAIN OVER THE RIGHT MANDIBLE, SUDDENLY APPEARING IN THE ACT OF SWALLOWING, MAY BE ASCRIBED TO:
  A. Cancer of the lower esophagus
  B. Cancer of the larynx
  C. Megaesophagus
  D. Tic douloureaux   Ref. p. 587

622. THE ABOVE DESCRIBED PAIN COULD BE SUPPRESSED BY SELECTIVE INTERRUPTION OF THE:
  A. Medial lemniscus
  B. Spinal tract of the trigeminal nerve
  C. Nucleus ambiguus
  D. Tractus solitarius   Ref. p. 587

623. A VIOLENT HEADACHE CAN BE CAUSED DURING AN OPERATION OF THE BRAIN, USING LOCAL ANESTHESIA, IF THE SURGEON INADVERTENTLY:
  A. Touches the occipital lobe
  B. Touches the frontal lobe
  C. Touches the somatic sensory cortex I
  D. Pulls on the tentorium (of the cerebellum)
                                  Ref. p. 587

624. A HEADACHE CAUSED BY A TUMOR IN THE PARIETAL LOBE REACHES THE NEURAL CENTERS OF PAIN THROUGH FIBERS CONTAINED IN THE:
  A. Intermediate nerve   D. Trigeminal nerve
  B. Facial nerve   E. Second cervical nerve
  C. Glossopharyngeus nerve   Ref. p. 587

625. AN INFRATENTORIAL TUMOR (CEREBELLAR) WHICH GIVES RISE TO A HEADACHE WILL SEND ITS PAINFUL IMPULSES TO THE PAIN CENTERS IN THE BRAIN THROUGH THE:
  A. Intermediate nerve   D. Trigeminal nerve
  B. Facial nerve   E. Second cervical nerve
  C. Glossopharyngeus nerve   Ref. p. 587

SECTION IV - PHYSIOLOGY OF THE NERVOUS SYSTEM

626. THE REMOVAL OF 50 ml OF CEREBROSPINAL FLUID IS FREQUENTLY ASSOCIATED WITH:
     A. A compensatory hypertension of the intracranial contents
     B. Cardiac arrhythmias
     C. A severe headache
     D. Dyspnea
     E. Tetany   Ref. p. 588

627. TRANSIENT LOCALIZED LOSS OF VISION (FOCAL SCOTOMATA):
     A. Is always associated with cerebrospinal hypertension
     B. Is invariably a sign of thrombosis of the retinal artery
     C. Is invariably a very ominous sign
     D. May precede a migraine attack   Ref. p. 588

628. THE REASON WHY A CERTAIN OBJECT CAUSES A SENSATION TO BE TERMED "HOT" WHILE ANOTHER OBJECT IS FELT "COLD" LIES IN THE FACT THAT:
     A. The frequency of discharge of neural impulses from all skin receptors sensitive to temperature is increased linearly with increasing temperature (frequency coding)
     B. The frequency of neural impulses from all skin receptors sensitive to temperature is decreased linearly with increasing temperature
     C. There are specific receptors connected to individual fibers firing optimally at cold temperatures while other separate specific receptors with their fibers fire optimally at warmer temperatures
     D. The sensation of "heat" is transmitted by mildly stimulated pain fibers while the sensation of "cold" is transmitted when all sensory input is transiently interrupted   Ref. p. 589

629. IF A TEMPERATURE OF 30^0 AT WHICH A "COLD" SKIN RECPTOR AND ITS NERVE ARE EXPOSED IS SUDDENLY DROPPED TO 28^0 C, ONE WOULD EXPECT FROM COMPARABLE EXPERIMENTS THAT SUCH RECEPTOR WOULD FIRE:
     A. More frequently without any adaptation at 28^0 C
     B. Less frequently without any adaptation at 28^0 C
     C. Less frequently with marked adaptation at 28^0 C
     D. More frequently with marked adaptation at 28^0 C
                                              Ref. Fig. 50-12, p. 590

630. BECAUSE OF THE PHYSIOLOGICAL BEHAVIOR OF TEMPERATURE-SENSITIVE RECEPTORS, THE SKIN OF A PERSON, SUDDENLY IMMERSED IN A POOL WITH THE TEMPERATURE OF THE WATER AT 20^0 C, WILL IN THE FIRST FEW MINUTES FEEL:
     A. Much colder than after 5 minutes of immersion
     B. Less cold than after 5 minutes of immersion
     C. Equally as cold as after 5 minutes of immersion
     D. Pain   Ref. p. 590

631. THE ADAPTATION OF COLD RECEPTORS IS KNOWN TO BE:
     A. Complete, to "extinction"
     B. Zero
     C. Partial and stable after an interval of time
     D. Constantly increasing   Ref. p. 590

SECTION IV - PHYSIOLOGY OF THE NERVOUS SYSTEM

632. TRANSMISSION OF ACTION POTENTIALS ORIGINATED IN THERMAL RECEPTORS IS BELIEVED TO TAKE PLACE IN FIBERS OF:
    A. Type A (alpha) 13-22 microns in diameter
    B. Type A (beta) 8-13 microns in diameter
    C. Type A (gamma) 4-8 microns in diameter
    D. Type A (delta) 1-4 microns in diameter
    Ref. p. 590

633. THE FORCE OF SOUND WAVES APPLIED ORIGINALLY TO THE TYMPANIC MEMBRANE IS AMPLIFIED BY THE RATIO
    <u>SURFACE AREA TYMPANIC MEMBRANE</u>
    SURFACE AREA OF STAPES
    AND THE INCREMENT IN FORCE ACCOMPLISHED BY THE FAVORABLE LEVERAGE CONTRIBUTED BY THE OSSICULAR SYSTEM. THIS INCREMENT IN FORCE EQUALS ABOUT:
    A. Two times
    B. Twenty-two times
    C. Fifty-two times
    D. One hundred and two times
    Ref. p. 592

634. THE NATURAL FREQUENCY OF THE OSSICLES OF THE MIDDLE EAR AND ITS LIGAMENTS AND THE NATURAL RESONANCE OF THE EXTERNAL EAR CANAL COMBINED, PROVIDE FOR AN OPTIMAL FREQUENCY RANGE FOR TRANSMISSION OF SOUND WAVES. THIS RANGE IS APPROXIMATELY BETWEEN:
    A. 1-500 cycles per second
    B. 600-6000 cycles per second
    C. 7000-15,000 cycles per second
    D. 15,000-50,000 cycles per second
    Ref. p. 593

635. A VERY LOUD SOUND IS REFLEXLY ATTENUATED BY SIMULTANEOUS CONTRACTION OF MUSCLES THAT EXERT A PULL ON THE HANDLE OF THE MALLEUS AND ON THE STAPEDIUS. THUS, IN SUCH A REFLEX THE HANDLE OF THE MALLEUS IS PULLED:
    A. Outward and the stapedius is pulled outward
    B. Outward and the stapedius is pulled inward
    C. Inward and the stapedius is pulled outward
    D. Inward and the stapedius is pulled inward
    Ref. p. 593

636. THE THREE PARALLEL JUXTAPOSED COILED TUBES THAT COMPOSE THE COCHLEA ARE ARRANGED IN SUCH A FASHION THAT THE SCALA:
    A. Vestibuli is separated from the scala media by the basilar membrane
    B. Vestibuli is separated from the scala tympani by the basilar membrane
    C. Media is separated from the scala tympani by the vestibular membrane
    D. Media is separated from the scala tympani by the basilar membrane
    Ref. Fig. 51-2, p. 594

SECTION IV - PHYSIOLOGY OF THE NERVOUS SYSTEM

THE FOLLOWING QUESTIONS OR INCOMPLETE STATEMENTS ARE FOLLOWED BY FOUR POSSIBLE ANSWERS. ANSWER BY USING THE KEY OUTLINED BELOW
A. If only 1, 2 and 3 are correct
B. If only 1 and 3 are correct
C. If only 2 and 4 are correct
D. If only 4 is correct
E. If all are correct

637. IT IS TRUE TO SAY THAT:
1. Sound vibrations are transmitted to the endolymph of the scala vestibuli through the stapes attached to the oval window
2. Fluid is moved in the scala vestibuli and the scala media almost at the same time
3. The helicotrema makes the scala vestibuli and the scala tympani a continous space
4. The basilar membrane oscillates up and down in response to pressure variations in the endolymph Ref. p. 594

638. RAPID VIBRATIONS OF THE STAPES CAUSE:
1. A rapid transmission of pressure-induced waves of endolymph to the round window through the helicotrema
2. An increased stiffening of the basilar membrane with high fidelity of transmission of endolymphatic movement to the scala media only
3. Obstruction of the helicotrema
4. The basilar membrane to vibrate because of the lag in fluid wave transmission through the helicotrema
Ref. p. 594

639. THE FIBERS WHICH COMPOSE THE BASEMENT MEMBRANE:
1. Are 12 times shorter in the helicotrema than at the base
2. Vibrate at a higher frequency at the base than at the apex of the cochlea
3. Move endolymphatic fluid with greater difficulty at the base of the cochlea than at the apex because of the difference in mass of fluid to be moved in these two regions
4. Have a "volume elasticity" which is 100 times greater at the helicotrema than at the base of the cochlea Ref. p. 594

640. VIBRATIONS OF THE BASILAR MEMBRANE DUE TO VIBRATIONS OF THE STAPES:
1. Are caused by endolymphatic fluid inertia
2. Are set in motion by the elasticity of fibers that compose the basilar membrane
3. Travel longitudinally in the direction of the helicotrema
4. Have a natural resonant frequency which is specific for every level of the cochlea Ref. p. 595

641. WHEN SOUND WAVES REACH THE LEVEL WHERE THE RESONANT FREQUENCY OF THE BASILAR MEMBRANE IS SIMILAR TO THE FREQUENCY OF THE SOUND:
1. The vibrations of the basilar membrane become maximal
2. There is no further propagation of vibrations beyond this point
3. The energy of the sound wave becomes rapidly dissipated beyond its resonant point
4. A large fraction of the wave's energy is propagated to the helicotrema in the case of high frequency sounds Ref. p. 595

642. IN THE PRODUCTION OF AUDITORY SENSORY IMPULSES:
1. The reticular lamina is rigidly continuous with the rods of Corti and the basilar membrane
2. The electrical potential of the rods of Corti initiate a receptor potential
3. Upward movement of the fibers of the basilar membrane lifts the reticular lamina and moves it in a shearing fashion inward
4. Hair cells membrane potential becomes strongly negative (hyperpolarized) while the rods of Corti initiate action potentials
   Ref. p. 596

643. THE ENDOCOCHLEAR POTENTIAL:
1. Varies rapidly when hair cells are moved by movements of the basilar membrane
2. Exists without variation at a level of about 80 mV, positive in the endolymph and negative in the perilymph
3. Is determined by the fact that the apex of hair cells is in contact with the endolymph while the base of these cells is in contact with the perilymph
4. Is caused by the continual secretion of K+ ions by the stria vascularis
   Ref. p. 597

644. THE CAPACITY TO PERCEIVE DIFFERENCES IN THE PITCH OF SOUNDS:
1. Is probably due to the selective resonant frequency of each level of the basilar membrane
2. Is dependent on the intact representation of cochlear nerve fibers in the cochlear nuclei
3. Is dependent on the spatial organization of fibers ascending in the brain stem
4. Depends on the somatotopic representation in the auditory fields of the cerebral cortex   Ref. p. 597

645. A SOUND IS PERCEIVED AS BEING MORE INTENSE THAN A PREVIOUS ONE OF THE SAME FREQUENCY BECAUSE THE SOUND WITH GREATER INTENSITY:
1. Causes the basilar membrane to vibrate with greater amplitude
2. Causes peripheral areas of the specific portion of the basilar membrane to vibrate, leading to "spatial summation" of impulses
3. Stimulates certain specific hair cells which are only stimulated by louder sounds
4. Stimulates pain receptors which summate with auditory impulses to reinforce the latter   Ref. p. 598

646. SOUND PERCEPTION INCREASES AS A FUNCTION OF THE INCREASE IN SOUND INTENSITY. IN ORDER TO OBTAIN A STRAIGHT LINE RELATIONSHIP ONE SHOULD PLOT THE SOUND PERCEPTION (ORDINATES) VERSUS THE:
1. Intensity of sound
2. Second power of the intensity of sound
3. Third power of the intensity of sound
4. Cube root of the intensity of sound   Ref. p. 598

SECTION IV - PHYSIOLOGY OF THE NERVOUS SYSTEM

SELECT THE ONE MOST APPROPRIATE ANSWER:

647. IF ONE MICROWATT OF SOUND ENERGY PER CENTIMETER OF SURFACE AREA IS USED AS ONE UNIT OF SOUND INTENSITY, IT WOULD BE TRUE TO SAY THAT A SOUND WITH A FREQUENCY OF 2,000 CYCLES PER SECOND WILL CAUSE IN A NORMAL HUMAN EAR A PERCEPTIBLE SOUND WHEN IT HAS AN INTENSITY EQUAL TO ABOUT:
    A. 0.1 microwatt/cm^2
    B. 0.001 microwatt/cm^2
    C. 0.000,000,1 microwatt/cm^2
    D. 1 microwatt/cm^2
    E. 10 microwatts/cm^2
    Ref. p. 598

648. USING THE SAME UNIT OF SOUND ENERGY (I_0) AS QUOTED IN THE PREVIOUS EXAMPLE (1 microwatt/cm^2) AND KNOWING THAT BY CONVENTION 1 bel = $\log \frac{I}{I_0}$ AND THAT 1 decibel (1 db) = 10 $\log \frac{I}{I_0}$ IT WOULD BE CORRECT TO STATE THAT THE SOUND OF MINIMAL INTENSITY WHICH THE HUMAN EAR MAY PERCEIVE HAS AN INTENSITY EQUAL TO:
    A. 1 decibel
    B. -30 decibels
    C. -50 decibels
    D. -70 decibels
    E. -90 decibels
    Ref. p. 598

649. IN ORDER TO EXPLORE THE COMPLETE RANGE OF THRESHOLD AUDIBILITY OF SOUNDS OF DIFFERENT FREQUENCIES BY THE HUMAN EAR, IT IS NECESSARY TO USE:
    A. A very low intensity of sounds throughout the whole spectrum of frequencies
    B. Ear phones
    C. High values of sound energies that may reach 10^{14} times the energy used at 2000 cycles per second
    D. A well trained subject
    E. All of the above
    Ref. p. 598

650. THE NEURONAL CHAIN CONDUCTING AUDITORY IMPULSES FROM THE SPIRAL GANGLION TO THE AUDITORY CEREBRAL CORTEX INCLUDES UP TO 6 SYNAPTIC JUNCTIONS AND NEURONS. NEURONS LOCATED IN THE DORSAL AND VENTRAL COCHLEAR NUCLEI HAVE AXONS WHICH TRANSMIT THEIR SIGNALS TO THE NEXT RELAY STATION, WHICH IS LOCATED IN THE:
    A. Inferior colliculi
    B. Medial geniculate nucleus
    C. Commissure of Probst
    D. Superior olivary nucleus
    E. Nucleus of the lateral lemniscus
    Ref. p. 599

651. A NEUROANATOMICAL FEATURE OF THE DISTRIBUTION OF AUDITORY PATHWAYS BECOMES OF GREAT VALUE IN PREVENTING DEAFNESS DUE TO UNILATERAL NEUROLOGICAL DISEASE. THIS CONSISTS OF THE FACT THAT:
    A. The superior olivary nucleus is very deep in the brain stem
    B. The trapezoid fibers are voluminous
    C. There are two cochlear nuclei: the dorsal and the ventral
    D. The pathways cross to the opposite side at three places (trapezoid fibers, commissure of Probst and commissure between inferior colliculli)
    E. There are many collaterals to the reticular activating system of the brain stem
    Ref. p. 599

SECTION IV - PHYSIOLOGY OF THE NERVOUS SYSTEM

652. THE RATE AT WHICH ACTION POTENTIALS ARE TRANSMITTED IN THE AUDITORY TRACTS DEPENDS VERY MARKEDLY ON SEVERAL FACTORS. IT IS TRUE THAT:
A. The frequency of firing in ascending auditory tracts, beyond the cochlear nuclei, is synchronous with the frequency of impulses arriving from the spiral ganglion
B. When no sound stimulates the organ of Corti there are no action potentials in the ascending pathways
C. Single nerve fibers originated in the spiral ganglion can transmit up to 1,000 impulses per second
D. The rate of cochlear nerve firing (of action potentials) is independent of the intensity (loudness) of the sound
Ref. p. 599

653. THE FINAL PROJECTION OF THE AUDITORY PATHWAYS TO THE CEREBRAL CORTEX TAKES PLACE IN THE AREA OF THE:
A. Geniculo-calcarine gyrus
B. Gyrus cinguli
C. Superior frontal gyrus
D. Supramarginal gyrus
E. Superior temporal gyrus
Ref. p. 600

654. THE AUDITORY CORTEX SEEMS CRITICALLY IMPORTANT TO RECOGNIZE:
A. Sounds of very low frequency
B. Sounds of very high frequency
C. Combination or sequence of tones
D. Sounds of very low intensity
Ref. p. 601

655. ASSUME A PERSON IS LOOKING TOWARD HIS LEFT SIDE AND THAT ANOTHER PERSON WITH A DEEP VOICE (LOW FREQUENCY SOUND) SPEAKS SOFTLY FROM HIS RIGHT SIDE. THE PERSON SPOKEN TO KNOWS ABOUT THE DIRECTION OF THE SOURCE OF SOUNDS OF LOW FREQUENCY BECAUSE THE:
A. Basilar membrane in the right ear is struck more forcibly than on the left
B. Sound waves reach the right ear earlier than at the left ear
C. Ipsilateral acoustic brain cortex is selectively activated while the contralateral side is inhibited
D. Superior colliculi on the right is stimulated to a greater extent than on the left side
Ref. p. 601

656. DETECTION OF THE DIRECTION OF THE SOURCE OF SOUNDS DEPENDS UPON COMPLEX MECHANISMS IN THE CHAIN OF ACOUSTIC NEURONS. THE SUPERIOR OLIVARY NUCLEI ARE:
A. Initially stimulated by ipsilateral sounds and inhibited by contralateral sounds
B. Initially depressed by ipsilateral sounds and stimulated by contralateral sounds
C. Simultaneously (on both sides) stimulated by sounds coming from the right
D. Simultaneously (on both sides) inhibited by sounds coming from the right
Ref. p. 601

SECTION IV - PHYSIOLOGY OF THE NERVOUS SYSTEM

657. CONDUCTION IN THE AUDITORY PATHWAYS:
 A. Can never occur from the olivary nuclei to the organ of Corti
 B. Is inhibited by the sense of smell
 C. Is not somatotopic
 D. May be retrograde and inhibit the organ of Corti's sensitivity
 Ref. p. 602

658. CONDUCTION OF A PURE TONE (TUNING FORK) NORMALLY:
 A. Lasts longer through vibration of the tympanic membrane and ossicular system (tuning fork held in front of the external ear canal) than through vibration of bony structures and endolymph (tuning fork held over the mastoid)
 B. Lasts longer through vibration of bony structures of the mastoid and endolymph (tuning fork held over the mastoid) than through vibration of the tympanic membrane and ossicular system
 C. Is carried to the right side preferentially when the tuning fork is held over the vertex of the calvarium
 D. Is suppressed by swallowing   Ref. p. 602

659. DEAFNESS DUE TO FUNCTIONAL IMPAIRMENT OF THE COCHLEA AND AUDITORY NERVE ON ONE SIDE WILL CAUSE:
 A. A selective loss of hearing of high frequency sounds by air conduction but good preservation of perception through bone conduction
 B. A similar degree of loss of hearing through air and bone conduction
 C. A painful reaction to sounds on the deaf side
 D. Auditory hypersensitivity to low frequency sounds propagated by bone conduction   Ref. p. 603

660. SCLEROSIS OF THE JOINTS OF THE OSSICULAR SYSTEM OF THE MIDDLE EAR MAY FOLLOW A CHRONIC INFECTION IN THIS AREA. THE MOST SEVERE DYSFUNCTION OCCURS WITH PERCEPTION OF:
 A. Low frequency sounds through air conduction
 B. High frequency sounds through air conduction
 C. Low frequency sounds through bone conduction
 D. High frequency sounds through bone conduction
 Ref. Fig. 51-12, p. 603

661. REFRACTION OF LIGHT RAYS AT AN INTERFACE OF TWO MEDIA WITH DIFFERENT INDICES OF REFRACTION (AIR/GLASS):
 A. Is a constant value for all degrees of angulation between the interface and the wave front of the light beam
 B. Increases with increments in the ratio of the two refractive indices
 C. Causes incident light that strikes the glass at $90°$ angle to deviate $90°$ from the incident light
 D. Is due to an increase in the velocity of propagation of light in the medium with a higher refractive index
 Ref. p. 604

SECTION IV - PHYSIOLOGY OF THE NERVOUS SYSTEM

662. THE REFRACTION OF PARALLEL LIGHT RAYS TAKES PLACE IN LENSES OF DIFFERENT SHAPES CAUSING VARIOUS TYPES OF IMAGES. FOCUSING OCCURS IN A:
   A. Cylindrical convex lens on the same side of the incident rays at a common focal point
   B. Cylindrical concave lens on the opposite side of the incident rays forming a focal line
   C. Spherical convex lens on the opposite side of the incident rays forming a focal point
   D. Spherical concave lens on the same side of the incident light forming a focal line
   Ref. p. 605

663. THE FOCAL DISTANCE OF A CONVEX SPHERICAL LENS IS:
   A. Longer for a point source in front of the lens than for a source of parallel rays placed at the same distance from the lens as the point source
   B. Shorter for a lens with a greater radius of curvature than for another with a shorter radius of curvature
   C. Is shorter for a lens with a smaller radius of curvature than for another with a longer radius of curvature
   D. Not influenced by the wavelength of the incident light
   Ref. p. 606

664. THE MATHEMATICAL RELATIONSHIP BETWEEN FOCAL DISTANCE (f), AND DISTANCE FROM POINT SOURCE TO LENS (a), AND DISTANCE FROM LENS TO FOCAL POINT (b) MAY BE EXPRESSED CORRECTLY AS FOLLOWS:
   A. $\frac{1}{f} = \frac{1}{a} + \frac{1}{b}$
   B. $f = a + b$
   C. $a = f + b$
   D. $\frac{1}{a} = \frac{1}{f} + \frac{1}{b}$
   E. $\frac{1}{b} = \frac{1}{a} + \frac{1}{f}$
   Ref. p. 606

665. IF THE REFRACTIVE POWER OF A LENS IS STATED TO BE + 5 DIOPTERS, ITS FOCAL LENGTH EQUALS:
   A. 0.20 meters
   B. 0.50 meters
   C. 1.00 meter
   D. 5.00 meters
   E. 10.00 meters
   Ref. p. 607

   Hint: f. length (meters) x diopters = 1 meter

SECTION IV - PHYSIOLOGY OF THE NERVOUS SYSTEM

666. THE REFRACTIVE POWER OF A LENS EXPRESSED IN DIOPTERS IS FUNCTIONALLY RELATED TO THE FOCAL LENGTH IN A MANNER THAT:
   A. Focal length is directly proportional to the diopters of the lens (with a slope equal to 1 meter)
   B. Diopters of the lens are inversely proportional to the focal length with a constant equal to 1 meter
   C. Diopters of the lens increase exponentially with increases in focal length
   D. Diopters of the lens increase logarithmically with increases in focal length
   Ref. p. 607

667. A CONCAVE SPHERICAL LENS WITH A DIOPTRIC POWER EQUAL TO -2 DIOPTERS:
   A. Diverges incident parallel light rays the same order of magnitude that a + 2 diopter lens would converge them
   B. Focuses the refracted rays at a point located 0.5 meters away from the lens on the opposite side of incoming rays
   C. Juxtaposed next to a + 2 dioptric spherical convex lens will have a focal point equal to 1 meter
   D. Gives rise to an upright image on the opposite side of the incident light
   Ref. p. 607

668. AMONG THE REFRACTIVE SURFACES OF THE EYE, THE HIGHEST REFRACTIVE INDEX IS THAT OF THE:
   A. Cornea
   B. Aqueous humor
   C. Crystalline lens
   D. Vitreous humor
   Ref. p. 608

669. THE MAJOR FACTOR WHICH CONTRIBUTES TO THE HIGH DIOPTRIC POWER CONTRIBUTED BY THE CORNEA IS DUE TO THE:
   A. Distance that separates it from the retina
   B. Fact that the cornea has a lower refractive index than the aqueous humor
   C. Lack of vascularity of the cornea
   D. Marked curvature of the anterior surface of the cornea
   E. Higher refractive index of the cornea compared to that of air
   Ref. p. 608

670. AS LIGHT RAYS PASS FROM THE CORNEA INTO THE AQUEOUS HUMOR THE REFRACTION OF LIGHT OCCURS IN A MANNER SIMILAR TO THAT TAKING PLACE THROUGH A:
   A. Convex cylindrical lens
   B. Spherical concave lens
   C. Spherical convex lens
   D. Neutral lens
   Ref. p. 608

671. THE EFFECT DESCRIBED IN THE PREVIOUS QUESTIONS IS DUE TO THE FACT THAT THE:
   A. Refractive index of the aqueous humor is lower than that of the cornea
   B. Refractive index of the aqueous humor is higher than that of the cornea
   C. Refractive index of the aqueous humor is the same as that of the cornea
   D. The aqueous humor can be optically considered to present the same features as air as far as light passage is concerned
   Ref. p. 608

SECTION IV - PHYSIOLOGY OF THE NERVOUS SYSTEM

672. THE PHYSICAL DESCRIPTION OF THE OPTICS OF THE HUMAN EYE ARE GREATLY SIMPLIFIED WHEN ALL REFRACTIVE SURFACES (CORNEA, AQUEOUS HUMOR, CRYSTALLINE LENS AND VITREOUS HUMOR) ARE ADDED TOGETHER AND CONSIDERED AS ONE LENS WHOSE CENTRAL POINT IS PLACED 17 mms IN FRONT OF THE RETINA. THIS IS KNOWN AS A "REDUCED EYE." KNOWING THAT DIOPTRIC POWER EQUALS THE RECIPROCAL OF THE FOCAL DISTANCE, WE MAY CALCULATE THAT ITS REFRACTIVE POWER EQUALS APPROXIMATELY:
    A. 1 diopter
    B. 9 diopters
    C. 20 diopters
    D. 59 diopters
    E. 109 diopters
    Ref. p. 608

    Hint: $D = \dfrac{1}{0.017 \text{ meters}}$

673. THE PHYSICAL ARRANGEMENT OF REFRACTIVE ELEMENTS IN THE EYE CAUSES THE RETINAL IMAGE TO BE:
    A. Invariably up-side down
    B. Out of focus
    C. Colored
    D. Occasionally larger than the object itself
    Ref. p. 608

674. THE PHENOMENON OF "ACCOMODATION" OF THE CRYSTALLINE LENS TO FOCUS ON A NEARER OBJECT IS DUE TO THE FACT THAT THE:
    A. Tension of the ligaments increases as a result of contraction of the ciliary muscle
    B. Lens approximates more closely its resting elasticity and becomes more spherical
    C. Insertion of the ligaments on the ciliary muscles are pulled backward and downward
    D. Sphincteric contraction of the ciliary muscle pulls further on the capsule of the crystalline
    Ref. p. 608

675. THE CILIARY MUSCLE IS INNERVATED AND CONTROLLED BY NERVES THAT REACH THE EYE WITH THE:
    A. Optic nerve
    B. Sympathetic fibers that reach the eye with the carotid plexus
    C. Third cranial nerve (parasympathetic fibers)
    D. Trochlear nerve
    Ref. p. 608

676. THE OPTICAL DEFECT THAT CHARACTERIZES THE CONDITION KNOWN AS PRESBYOPIA CONSISTS OF A(N):
    A. Permanent fixation of the crystalline lens in the most powerful dioptric state it can assume
    B. Elongation of the distance from the lens to the retina in the reduced eye
    C. Shortening of the distance from the lens to the retina in the reduced eye
    D. Loss of the capacity of the lens to assume a more powerful dioptric position
    Ref. p. 609

SECTION IV - PHYSIOLOGY OF THE NERVOUS SYSTEM

677. THE RADIUS OF THE HUMAN PUPIL MAY VARY FROM 0.75 MMS WHEN MAXIMALLY CONSTRICTED TO ABOUT 4 MMS, AN INCREASE EQUAL TO ABOUT 5.3 TIMES, WHEN MAXIMALLY DILATED. THE AMOUNT OF LIGHT THAT PASSES THROUGH THE PUPIL IS A FUNCTION OF ITS AREA, AND NOT OF ITS RADIUS ALONE. THUS THE INCREMENT IN LIGHT REACHING THE RETINA FROM A MAXIMAL MIOSIS TO A MAXIMAL MYDRIASIS EQUALS ABOUT:
- A. 5.0 times
- B. 10.0 times
- C. 30.0 times
- D. 90.0 times
- E. 150 times
Ref. p. 609

Hint: Increment in area = $\dfrac{\pi (4)^2}{\pi (0.75)^2}$

678. WHEN THE PUPIL IS MAXIMALLY CONSTRICTED BY A MYOTIC AGENT:
- A. The retinal image may disappear altogether
- B. There is a tendency to cause myopia due to an elongation of the distance between lens and retina
- C. The depth of focus is maximal
- D. The depth of focus is minimal
- E. The depth of focus undergoes no change
Ref. p. 609

679. A MAXIMAL PUPILLARY CONSTRICTION IS MOST LIKELY TO INCREASE THE CHANCE OF INDUCING A GREATER DEGREE OF:
- A. Spherical aberration
- B. Chromatic aberration
- C. Diffractive errors
- D. Myopia
Ref. p. 610

680. A MAJOR REQUIREMENT FOR THE CORRECT DIAGNOSIS OF AN ERROR OF REFRACTION IS THAT THE:
- A. Pupil must be maximally dilated
- B. Ciliary muscle must be able to contract under its parasympathetic influence
- C. Lens must have a normal elasticity
- D. Ciliary muscle must be completely relaxed
Ref. p. 610

681. HYPERMETROPIA IS CHARACTERIZED BY A(N):
- A. Retinal image which tends to form in front of the retina when the ciliary muscle is relaxed
- B. Very long anteroposterior eye diameter
- C. Improvement of the condition with the development of presbyopia
- D. Permanent state of contraction of the ciliary muscle which tends to obtain a more clear retinal image by an increase in the dioptric power of the crystalline lens
Ref. p. 611

132   SECTION IV - PHYSIOLOGY OF THE NERVOUS SYSTEM

682. THE MYOPIC PATIENT WHO HAS A SEVERE CASE OF THE RE-
FRACTIVE ERROR:
A. May obtain a clear retinal image by voluntary adjustment of the con-
tractile tone of his ciliary muscle
B. Is unable to form clear retinal images unless a negative lens de-
creases the dioptric power of his eye
C. Has a deterioration of his refractive defect with the development of
presbyopia
D. May see his defect corrected by the use of a convex spherical lens
Ref. p. 611

683. IN AN ASTIGMATIC PATIENT:
A. Complete correction of the refractive error may be obtained with a
spherical lens of the proper dioptric power
B. Two cylindrical lenses of equal strength may be crossed at right
angles to correct the refractive defect
C. The light rays in one plane (vertical) have focal points placed at a
different retinal level from focal points that correspond to another
plane perpendicular to the first (horizontal)
D. The refractive defect may be corrected by a conscious application of
the accomodative function of the crystalline lens
Ref. p. 612

684. IN A NORMAL EYE, THE RETINAL IMAGE PRODUCED BY SEEING A
CHILD WHOSE HEIGHT EQUALS 1 METER (3 FEET, 3 INCHES), STAND-
ING AT A DISTANCE EQUAL TO 17 METERS IN FRONT OF THE EYE
(ABOUT 56 FEET) EQUALS APPROXIMATELY:
A. 100 microns
B. 1 millimeter
C. 10 millimeters
D. 50 millimeters
E. 100 millimeters   Ref. p. 612

Hint:
                                              Ref.pg. 612

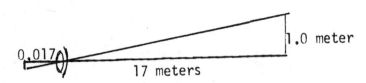

SECTION IV - PHYSIOLOGY OF THE NERVOUS SYSTEM

685. VISUAL ACUITY IS MEASURED BY THE MINIMAL ANGLE THAT TWO INCIDENT RAYS OF LIGHT FORM AS THEY APPROACH THE EYE AND WHICH CAN BE RECOGNIZED AS ORIGINATING IN TWO DISTINCT LIGHT SOURCES. THE VISUAL CORTEX IS BELIEVED CAPABLE OF RECOGNIZING TWO DISTINCT RETINAL IMAGES WHEN ONE IMAGE STIMULATES ONE CONE IN THE FOVEA AND THE OTHER IMAGE FALLS NOT IN THE CONE ADJACENT TO THE FIRST, BUT, AT LEAST ON A THIRD CONE, SO THAT THERE IS A MINIMAL OF ONE CONE WHICH IS NOT STIMULATED. THIS LATTER UNSTIMULATED CONE "SEES" THE SEPARATION BETWEEN THE POINT SOURCES. ASSUMING A MEAN CONE DIAMETER EQUAL TO 1.5 MICRONS, TWO POINT SOURCES PLACED AT A DISTANCE EQUAL TO 20 FEET (6.096 METERS) IN FRONT OF A SUBJECT, AND AN ANTEROPOSTERIOR EYE DIAMETER EQUAL TO 17 MILLIMETERS, THE MINIMAL SEPARATION THAT SHOULD EXIST BETWEEN THEM AT SUCH A DISTANCE EQUALS APPROXIMATELY:
    A. 0.5 millimeters
    B. 5.0 millimeters
    C. 50 millimeters
    D. 70 millimeters
    E. 100 millimeters   Ref. p. 613

Hint:

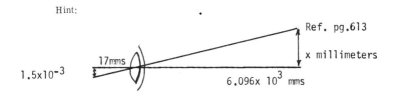

686. IF AN OPTHALMOLOGIC EVALUATION CONCLUDES THAT A SUBJECT HAS A $\frac{20}{200}$ VISUAL ACUITY, IT MEANS THAT AT A DISTANCE OF:
    A. 20 feet, the subject can see 200 times better than the normal
    B. 20 feet, the subject can see 200 times less than normal
    C. 20 feet, the subject can see what a normal person could see at 200 feet
    D. 200 feet, the subject could see what a normal person would see at 20 feet   Ref. p. 613

687. BINOCULAR PARALLAX (STEREOPSIS):
    A. Allows a person to judge the relative distance between two objects located at a far distance from the eyes
    B. Causes double vision
    C. Causes two separate objects to be seen as one by the formation of one retinal image
    D. Allows a person to have a telescopic view of objects
    E. Allows a person to judge the relative distance between two objects placed close to the subject   Ref. p. 613

SECTION IV - PHYSIOLOGY OF THE NERVOUS SYSTEM

688. A LIGHT SOURCE WHICH ILLUMINATES THE RETINA OF A SUBJECT ALLOWS THE EXAMINATION OF ANATOMICAL STRUCTURES IN THE DEPTH OF THE EYE. SUCH LIGHTED STRUCTURES IN THE RETINA EMERGE FROM THE EYE OF THE SUBJECT TOWARD THE EXAMINER THROUGH A PROCESS OF REFRACTION AS:
   A. Diverging rays
   B. Converging rays
   C. Parallel rays
   D. Collimated rays   Ref. p. 614

689. THE MACULA OF THE HUMAN RETINA:
   A. Contains the largest cones to be found in the retina
   B. Has cones and rods in about equal proportion
   C. Contains only rods in the fovea centralis
   D. Has a very thin layer of small cones with no ganglion cells in the fovea
                                               Ref. p. 617

690. ALBINOS ARE KNOWN TO HAVE A(N):
   A. High degree of visual acuity
   B. Lesser amount of light that can pass through the iris
   C. Decreased number of rods and cones excited by light
   D. Overreflectance of the choroid and retina
                                               Ref. p. 617

691. THE CHOROID OF THE EYE:
   A. Contains throughout its whole area the vascular network which perfuses the inner layers of the retina (ganglion cell layer)
   B. Is not directly involved with the vascular supply of rods and cones of the retina
   C. Stores vitamin A
   D. Will, if partially torn, cause an inevitable ischemic damage to the retina in all cases of retinal detachment
                                               Ref. p. 617

692. THE VISUAL PIGMENT CONTAINED WITHIN THE RODS:
   A. Is called scotopsin
   B. Is free of protein when it is in pure form
   C. Undergoes a change in the spatial position of a portion of the retinene molecule when light is absorbed
   D. Becomes reconstituted to the cis form of retinene before the electrical potential can be generated   Ref. p. 617

693. THE RECEPTOR POTENTIAL GENERATED IN RETINAL RODS WHICH HAVE BEEN EXPOSED TO THE ACTION OF LIGHT:
   A. Does not increase in magnitude when the intensity of light increases. It follows an "all or none" response
   B. Is biphasic and increases logarithmically with increases in the intensity of the stimulation in the second phase
   C. Has a much higher voltage in the first phase of response to one brief light stimulus
   D. Essentially has the same polarity as potentials obtained in other receptors   Ref. p. 618

SECTION IV - PHYSIOLOGY OF THE NERVOUS SYSTEM

694. THE MAJOR PART OF THE VITAMIN A WHICH CONTRIBUTES TO THE RECONSTITUTION OF THE ORIGINAL FORM OF RETINENE IN RHODOPSIN IS:
    A. Stored in the rods
    B. Stored in the cones
    C. Circulating free in blood perfusing the retina
    D. Stored in the pigment layer   Ref. p. 618

695. THE CHEMICAL CONVERSION OF VITAMIN A TO RETINENE INVOLVES THE:
    A. Transfer of hydrogen atoms to vitamin A
    B. Transfer of hydroxyl groups to vitamin A
    C. Removal of hydrogen atoms from vitamin A
    D. Removal of hydroxyl groups from vitamin A
    E. Transfer of methyl groups to vitamin A
                                            Ref. p. 619

696. AS A RESULT OF THE ABOVE DESCRIBED REACTION THE TERMINAL GROUP OF VITAMIN A IS CONVERTED FROM A(N):
    A. Alcohol to an aldehyde
    B. Acid to an aldehyde
    C. Aldehyde to an alcohol
    D. Carboxylic acid to an amide
    E. Phosphate group to a choline group   Ref. p. 619

697. THE FASTEST OF THE CHEMICAL REACTIONS THAT FOLLOW LIGHT-INDUCED CHANGES IN THE RETINAL RODS IS THE CONVERSION OF:
    A. Retinene into vitamin A
    B. Vitamin A into retinene
    C. Rhodopsin into retinene
    D. Scotopsin into rhodopsin   Ref. p. 619

698. WHEN THE RETINA IS EXPOSED TO INTENSE LIGHT FOR A LONG TIME, THE BULK OF PHOTOCHEMICALS AND INTERMEDIATES ARE IN THE FORM OF:
    A. Rhodopsin
    B. Retinene
    C. Vitamin A
    D. DPNH   Ref. p. 619

699. AFTER PROLONGED EXPOSURE TO A DARK ENVIRONMENT, THERE IS IN THE RETINA A NEGLIGIBLE AMOUNT OF:
    A. Rhodopsin
    B. Retinene
    C. DPN
    D. DPNH   Ref. p. 619

700. NIGHT BLINDNESS:
    A. May occur after a few days of vitamin A deficiency
    B. May be cured within a few hours of treatment with vitamin A
    C. Requires months of treatment before the functional capacity of retinal rods is improved
    D. Is due to diminished rate of conversion of the rich stores of retinene into the complete molecule of rhodopsin
                                            Ref. p. 619

SECTION IV - PHYSIOLOGY OF THE NERVOUS SYSTEM

701. THE NATURE OF THE LIGHT SENSITIVE CHEMICALS FOUND IN CONES DIFFERS FROM THAT FOUND IN RODS MAINLY DUE TO A DIFFERENT:
   A. Structure of the retinene fraction
   B. Concentration of rhodopsin per cell
   C. Arrangement of the chemicals in the photo-sensitive cell
   D. Chemical structure of the protein component of the photopsins
   Ref. p. 619

702. A PLOT OF LIGHT ABSORPTION OF RETINAL PIGMENTS OBTAINED FROM THE RETINA AS A FUNCTION OF THE WAVELENGTH REVEALS THAT RHODOPSIN:
   A. Has a peak absorption similar and superimposed to that which results from green-sensitive cones
   B. Absorbs maximally at 505 millimicrons
   C. Absorbs uniformly throughout the visible spectrum
   D. Is maximally sensitive to orange   Ref. p. 619

703. THE SENSITIVITY OF RETINAL RODS TO LIGHT IS OF SUCH A MAGNITUDE THAT:
   A. A decrease of 50% of the content of rhodopsin will result in 50% decrease in sensitivity
   B. A decrease of half a percent in the content of rhodopsin causes several thousand fold decrease in sensitivity
   C. A total exhaustion of the content of rhodopsin must occur before rods cease to respond maximally to a minimal light stimulus
   D. Breakdown of rhodopsin due to light stimuli actually causes an increase in the sensitivity to light   Ref. p. 620

704. WHEN A PERSON ENTERS A TOTALLY DARK ROOM, THE RETINAL SENSITIVITY TO LIGHT INCREASES IN THE FIRST TWENTY MINUTES (DURING THE PROCESS KNOWN AS DARK ADAPTATION) TO AN EXTENT EQUAL TO APPROXIMATELY:
   A. 5,000 fold
   B. 500 fold
   C. 50 fold
   D. 5 fold
   E. 2 fold
   Ref. Fig. 53-4, p. 620

705. BECAUSE OF DIFFERENCES IN RATES OF CHEMICAL CHANGES IN THE RETINA, THE ADAPTATION TO DARK OCCURS:
   A. More rapidly in cones
   B. More rapidly in rods
   C. Quantitatively to a greater extent in cones
   D. To the same extent (quantitatively) in rods as in cones
   Ref. p. 620

706. THE FASTEST ADAPTATION TO A CHANGE IN ILLUMINATION TAKES PLACE WHEN THE EYE HAS BEEN EXPOSED TO:
   A. Intense light for a few minutes and then is exposed to a dark environment (rate of dark adaptation)
   B. Intense light for one hour and then is exposed to a dark environment (rate of dark adaptation)
   C. Intense light for several hours and then is exposed to a dark environment (rate of dark adaptation)
   D. A dark environment and then is exposed to a lighted one (rate of light adaptation)   Ref. Fig. 53-5, p. 620

SECTION IV - PHYSIOLOGY OF THE NERVOUS SYSTEM

707. THE FREQUENCY AT WHICH FLICKER-FUSION TAKES PLACE INCREASES AS A FUNCTION OF THE INTENSITY OF ILLUMINATION BECAUSE:
   A. Cones have a lower excitability than rods but detect rapid changes in the intensity of illumination
   B. Rods have a higher excitability than cones and detect rapid changes of illumination
   C. Cones have a lower threshold of excitability than rods but detect less rapid changes in the intensity of illumination
   D. Rods have a higher threshold of excitability than cones but detect more slowly the changes in the intensity of illumination
   Ref. p. 622

708. IT IS BELIEVED THAT THERE ARE THREE DIFFERENT RETINAL CONES, EACH CAPABLE OF RESPONDING SELECTIVELY TO ONE COLOR IN THE VISIBLE SPECTRUM. ALL OF THE FOLLOWING COLORS HAVE A SPECIFIC CONE, EXCEPT:
   A. Blue
   B. Green
   C. Yellow
   D. Red
   Ref. p. 622

709. THE SENSATION OF RED COLOR TAKES PLACE IN THE BRAIN PROBABLY BECAUSE THE STIMULATION OF RETINAL CONES OCCURS IN SUCH A MANNER THAT BLUE:
   A. Green and red cones are equally stimulated
   B. Cones are not stimulated at all, while green cones are minimally and red cones are maximally stimulated
   C. Cones are not stimulated while the green and red are stimulated to the same extent
   D. And red cones are stimulated to the same extent while green cones are not
   Ref. p. 622

710. THE PERSON WHO CONGENITALLY IS INCAPABLE OF RECOGNIZING A RED LIGHT:
   A. Is a deuteranope
   B. Will "see" red colors with cones maximally excitable by green color (green cones) and interpret them as being "green"
   C. Has a minimal threshold around 500 millimicrons (wavelength light stimulus)
   D. Lacks the proper gene in one of the autosomes
   E. Is more commonly a female rather than a male
   Ref. p. 624

711. DIRECT SYNAPSES OF FIBERS PRESENT IN THE OPTIC NERVE OCCUR AT THE FOLLOWING SITES, EXCEPT THE:
   A. Lateral geniculate body
   B. Superior colliculi
   C. Pretectal nuclei of the brain stem
   D. Medial geniculate body
   E. Amacrine cells of the retina
   Ref. p. 626

SECTION IV - PHYSIOLOGY OF THE NERVOUS SYSTEM

712. THE PORTION OF THE RETINA WHICH COVERS THE PERIPHERAL AREAS OF THE EYE HAS SEVERAL DIFFERENCES FROM THAT WHICH COVERS THE FOVEA CENTRALIS. AN IMPORTANT PHYSIOLOGICAL DIFFERENCE CONSISTS OF THE GREATER SENSITIVITY OF THE PERIPHERAL RETINA TO WEAK LIGHT. THIS HAS BEEN ASCRIBED TO THE FACT THAT THE PERIPHERAL RETINA HAS A UNIQUE:
    A. Large type of rods with greater concentration of rhodopsin
    B. Large surface area
    C. Large number of cones
    D. Convergence of many rods on one ganglion cell
    E. Distribution in the calcarine gyrus   Ref. p. 627

713. THE LIGHT-INDUCED ELECTRICAL CHANGE IN THE OUTER SEGMENT OF RODS AND CONES IS CHARACTERIZED BY ALL OF THE FOLLOWING FEATURES, EXCEPT IT:
    A. Increases logarithmically with increments in the intensity of light stimulus
    B. Is transmitted unchanged through the body of cones
    C. Is transmitted unchanged through the body of rods
    D. Is an action potential   Ref. p. 627

714. GANGLION CELLS IN THE RETINA ARE KNOWN TO:
    A. Increase their frequency of discharge at a constant rate while light stimulates the corresponding rods and cones
    B. Transmit nerve impulses at a rate of 55 per second even in the dark
    C. Respond to the suppression of light by decreasing their rate of discharge in all cases
    D. Cause lateral inhibition through synapses on horizontal cells
    E. Lack the capacity to determine a contrast between light and dark areas of vision   Ref. p. 628

715. "CENTRIFUGAL" FIBERS WHICH SEEM TO BRING NEURAL MESSAGES FROM THE BRAIN STEM TO THE RETINA ESTABLISH DIRECT SYNAPTIC CONNECTIONS WITH:
    A. Ganglion cells   D. Horizontal cells
    B. Rods   E. Amacrine cells
    C. Cones   Ref. p. 629

716. NEURONS IN THE LATERAL GENICULATE BODY:
    A. Transmit messages from both nasal and temporal halves of the ipsilateral eye
    B. Are devoid of an "on-off" cycle of discharge coincidental with retinal events occurring at the same time
    C. Are layered so that the gradual inward distribution of neurons corresponds successively to the temporal (outer) and nasal (inner) halves of the retina
    D. Match somatotopically equivalent areas of the retina with the same pattern of "on-off" with lateral inhibition that the retina exhibits when stimulated by light   Ref. p. 629

SECTION IV - PHYSIOLOGY OF THE NERVOUS SYSTEM

717. THE CORTICAL REPRESENTATION OF HALF OF THE MACULA (IN THE RETINA) HAS THE FOLLOWING FEATURES, EXCEPT IT:
   A. Is located deep in the calcarine fissue at about 5 cms from the occipital pole
   B. Is surrounded by representation of more peripheral portions of the retina
   C. Is proportionally larger than other areas of the retina
   D. Contains in its midportion the area that represents the fovea centralis
   Ref. p. 630

718. THE STIMULATION OF THE VISUAL CORTEX IS:
   A. Very intense if a uniform white plane surface is the subject visualized
   B. Most effective in areas of high contrast in the visual scene
   C. Less effective when recognizing inclined contrasts than horizontal or vertical ones
   D. Qualitatively different from that in the geniculate body in that the latter does not recognize boundaries   Ref. p. 630

719. THE VISUAL ASSOCIATION AREAS ARE:
   A. Capable of responding to geometric patterns, such as right angles, lines, etc.
   B. Located in the tegmentum of the midbrain, next to the medial longitudinal fasciculus
   C. Capable of giving rise to visual perception even if the cortex in area 17 is destroyed
   D. Not related to color perception   Ref. p. 632

720. ALEXIA:
   A. May result from an isolated destruction of area 17
   B. Consists of the lack of visual perception of colored images
   C. Is the name given to the visual aura that precedes some epileptic seizures
   D. Consists of the visualization of letters whose meaning becomes incomprehensible   Ref. p. 632

721. THE MORE COMPLEX MENTAL PROCESS INVOLVED IN THE SEQUENCE THAT BEGINS WITH THE VISUALIZATION OF AN OBJECT AND THE MOVEMENTS WHICH LOGICALLY RESULT FROM SUCH VISUAL PERCEPTION IS LOCATED IN THE BRAIN IN THE AREA:
   A. Located immediately rostral to area 17 (area 18)
   B. Located in the medial portion of the motor cortex (area 4)
   C. Of the angular gyrus
   D. Of the cyngulate gyrus
   E. Of the limbic gyrus   Ref. p. 632

722. BECAUSE OF THE ANATOMICAL LOCATION OF THE ENTRANCE OF THE OPTIC NERVE WITH RESPECT TO THE EYEBALL, THE BLIND SPOT IS LOCATED:
   A. In the nasal portion of the visual field of the corresponding eye
   B. In the macular area of the opposite eye
   C. In the temporal visual field, at about $90°$ to the side of the center
   D. About $15°$ of the central point of vision, in the temporal field
   Ref. p. 632

SECTION IV - PHYSIOLOGY OF THE NERVOUS SYSTEM

723. RETINITIS PIGMENTOSA:
   A. Causes macular blindness first and peripheral blindness later
   B. Is an inflammation of the retina due to an abnormal metabolism of rhodopsin (the visual pigment present in rods)
   C. May be diagnosed by finding blind spots in a perimetry chart
   D. Is due to lead poisoning   Ref. p. 632

724. BITEMPORAL HEMIANOPSIA:
   A. May result from a lesion of the temporal retina of both sides
   B. May be produced if the mid-portion of the optic chiasma is compressed
   C. Consists of a longitudinal scotoma which cuts in half the temporal visual field
   D. Consists of an imbalance of vision between the temporal retinas on both sides   Ref. p. 633

725. THE TOTAL TRANSECTION OF THE OPTIC TRACT OF THE RIGHT SIDE:
   A. Causes blindness of the right eye
   B. Causes blindness of the left eye
   C. Gives rise to bi-temporal hemianopsia
   D. Causes blindness in the right temporal and left nasal visual fields
   E. Causes blindness on the left temporal and right nasal visual fields
                 Ref. p. 633

726. THE TRANSECTION OF ALL THE VISUAL RADIATIONS WHICH EMERGE FROM THE RIGHT LATERAL GENICULATE BODY AND TERMINATE IN THE RIGHT VISUAL CORTEX CAUSES:
   A. Total blindness of the right eye
   B. Total blindness of the left eye
   C. A visual defect almost identical to that obtained by transection of the right optic tract
   D. Numerous scotomata on the whole peripheral visual field of the right eye   Ref. p. 633

727. A LESION IN THE RIGHT OPTIC TRACT MAY BE DIFFERENTIATED FROM ANOTHER LESION IN THE RIGHT OPTICAL RADIATIONS BECAUSE A RIGHT OPTICAL TRACT LESION CAUSES:
   A. Color blindness on the right side, while a lesion of the right optical radiation does not
   B. Homonymous hemianopsia on the right side, while a lesion of the right optical radiation does not
   C. Nystagmus on the right side, while a lesion of the right optical radiation does not
   D. Loss of the pupillary reflex to light while a lesion of the right optical radiation does not   Ref. p. 633

728. THE RIGHT SUPERIOR OBLIQUE (EXTRAOCULAR MUSCLE OF THE EYE) IS INNERVATED BY:
   A. Fibers that originate in the right superior colliculi
   B. The IV cranial nerve on the right side
   C. The trochlear nerve on the left side
   D. The III cranial nerve on the left side
   E. The trigeminal nerve on the right side
                 Ref. Fig. 54-9, p. 633

SECTION IV - PHYSIOLOGY OF THE NERVOUS SYSTEM   141

729. CONJUGATE MOVEMENTS OF BOTH EYES WOULD BE AFFECTED
ADVERSELY IF A BRAIN TUMOR IS LOCATED IN THE AREA OF THE:
A. Vermis of the cerebellum   D. Superior colliculi
B. Medulla oblongata   E. Optic chiasm
C. Ponto-cerebellar angle   Ref. p. 633

730. A PERSON WHO IS ASKED TO LOOK TO HIS RIGHT SIDE AND IS UNABLE
TO DO SO (WHILE ABLE TO LOOK TO HIS LEFT SIDE) MAY HAVE A
LESION MOST LIKELY LOCATED IN THE:
A. Premotor (frontal lobe) area on the left side
B. Superior colliculi
C. Inferior colliculi
D. Superior temporal gyrus in the left side
E. Sensory-motor cortex on the right side
Ref. Fig. 54-7, p. 634

731. IMAGES FORMED IN THE FOVEA HAVE A TENDENCY TO REMAIN
THERE DUE TO CONTINUOUS RAPID REFLEX MOVEMENTS OF THE
EYE, AND ARE GREATLY DEPENDENT ON THE INTACT FUNCTION
OF THE:
A. Area 17 of the visual cortex
B. Frontal lobe
C. Motor cortex
D. III cranial nerve
E. Area 19 of the occipital cortex   Ref. p. 634

732. IN THE FORM OF NYSTAGMUS KNOWN AS OPTIKOKINETIC THE SLOW
COMPONENT OF THE EYE MOVEMENT:
A. Has an upward direction
B. Has a downward direction
C. Has a circular direction around the horizon
D. Follows the same direction of movement of a moving object in the
visual field
E. Returns the visual field to a new field in the same direction where
the movement is coming from   Ref. p. 635

733. IN A CASE OF STRABISMUS, THE EYE WHICH IS LEAST USED FOR
FIXATION ON THE OBJECT UNDER ATTENTION HAS:
A. A visual acuity of the same degree as the (normal) fixating eye
B. An atrophic nerve
C. Optikokinetic nystagmus as a rule
D. A marked diminution of visual acuity
E. No color vision   Ref. p. 636

734. THE PREGANGLIONIC SYMPATHETIC NERVES THAT REACH THE EYE
HAVE THEIR NEURONS LOCATED IN THE:
A. Superior colliculi
B. Nucleus of the VII cranial nerve (facial nerve)
C. Geniculate ganglion
D. Ciliary ganglion
E. Thoracic segment of the spinal cord   Ref. p. 636

SECTION IV - PHYSIOLOGY OF THE NERVOUS SYSTEM

735. THE PARASYMPATHETIC POST-GANGLIONIC FIBERS HAVE THEIR NEURONS LOCATED IN THE:
A. Edinger-Westphal nuclei
B. Ciliary ganglion
C. Geniculate ganglion
D. Nuclei of the VII cranial nerve (facial nerve)
E. Thoracic segment of the spinal cord   Ref. p. 636

736. A CONSTANT REGULATION OF THE DIOPTRIC POWER OF THE LENS IS REQUIRED FOR ACCOMMODATION TO SUCCEED IN FORMING SHARP IMAGES IN THE FOVEA. PRESUMABLY THIS DEPENDS ON AN INTACT CHAIN OF NEURONS WHICH HAVE THEIR REPRESENTATION IN:
A. Areas 18 and 19 of the occipital lobe
B. Area 17 of the occipital cortex (striate area)
C. Premotor area of the frontal lobe
D. The vestibular nuclei
E. Superior temporal gyrus   Ref. p. 637

737. THE NERVE FIBERS WHICH EMERGE FROM THE CENTRAL NERVOUS SYSTEM TO REACH THE EYE AND REGULATE ACCOMMODATION HAVE THEIR CELL BODIES (NEURONAL BODY) LOCATED IN THE:
A. Areas 18 and 19 of the occipital lobe
B. Vestibular nuclei
C. Premotor area of the frontal lobe
D. Edinger-Westphal nuclei (superior colliculi)
E. Nuclei of the IV cranial nerve   Ref. p. 637

738. PUPILLARY CONSTRICTION IS THE REFLEX RESPONSE OF THE EYE TO AN INCREASED INTENSITY OF LIGHT ENTERING THE EYE AND STIMULATING THE RETINA. THE INTACT REFLEX SEEN IN NORMAL HUMANS MAY BE ABOLISHED IF THERE IS NEURONAL DAMAGE OF THE NUCLEUS OF:
A. Edinger-Westphal
B. Trochlear nerve
C. Abducens nerve
D. Area 17 of the occipital cortex
E. Areas 18 and 19 of the occipital cortex
    Ref. p. 637

739. A CONCOMITANT PHENOMENON TAKING PLACE IN A NORMAL EYE WHICH ACCOMMODATES TO SEE AN OBJECT PLACED AT A SHORT DISTANCE (50 CMS) FROM THE EYES IS:
A. Pupillary dilatation
B. Pupillary constriction
C. Divergence of the eyes
D. A shortening of the depth of focus
E. A decreased sensitivity to color vision
    Ref. p. 637

SECTION IV - PHYSIOLOGY OF THE NERVOUS SYSTEM

740. IN SYPHILIS OF THE CENTRAL NERVOUS SYSTEM THE PUPILLARY CONSTRICTION THAT NORMALLY SHOULD FOLLOW THE EXPOSURE TO A BRIGHTER LIGHT DOES NOT OCCUR. THE PUPILS IN PATIENTS AFFLICTED BY THIS DISEASE:
   A. Do not constrict when the eyes converge to look at a closer distance
   B. Do not constrict because the parasympathetic fibers to the eye are destroyed by the lesions
   C. Do not constrict during accommodation to look at a closer distance
   D. Do not constrict because the sympathetic fibers to the eye are destroyed by the lesions
   E. Constrict during accommodation to look at a closer distance
   Ref. p. 637

741. THE ARGYLL-ROBERTSON PUPIL:
   A. Is incapable of constriction with myotics
   B. Is incapable of dilation with mydriatics
   C. Is mydriatic, with a fixed large pupillary diameter
   D. Is myotic, with a fixed small pupillary diameter
   E. Prevents light from passing through because of opacification in the lens
   Ref. p. 638

742. BECAUSE OF THE ANATOMICAL DISTRIBUTION OF THE SYMPATHETIC FIBERS OF THE EYE, THE SO-CALLED HORNER'S SYNDROME MAY BE FOUND IF:
   A. A metastatic tumor of the lungs compresses the area of the first thoracic vertebra
   B. Exophthalmos is due to hyperthyroidism
   C. The brain stem is compressed at the level of the superior colliculus
   D. The occipital lobes are infarcted by occlusion of the posterior cerebral arteries
   Ref. p. 638

743. THE SENSATION OF TASTE:
   A. Does not start with a "receptor potential" but with an action potential, the magnitude of which is proportional to the logarithm of the concentration of the stimulus
   B. Is conducted through axons that reach their final destination in the thalamus
   C. Requires an intact chain of 3 neurons: one in a peripheral ganglion, a second in the tractus solitarius and a third in the thalamus
   D. Is the only sensation not requiring an intact cerebral cortex for its perception
   E. Enters the brain stem through the chorda tympani only, with exclusion of other cranial nerves
   Ref. p. 642

744. STIMULATION OF THE OLFACTORY MUCOSA BY AN APPROPRIATE VOLATILE SUBSTANCE:
   A. Causes an excitatory state that does not undergo any degree of adaptation, provided the stimulus is continuous
   B. Gives rise to an electrical potential on the surface of the mucosa which is of the "all or none" type
   C. Causes a generator potential, the amplitude of which increases with the logarithm of the strength of the stimulus
   D. Suppresses the basal rate of discharge through olfactory nerves and reduces it to zero
   Ref. p. 644

SECTION IV - PHYSIOLOGY OF THE NERVOUS SYSTEM

745. MOST OF THE FIBERS OF THE CORTICOSPINAL TRACT HAVE DIRECT SYNAPTIC CONTACT WITH:
  A. Anterior motoneurons
  B. Sympathetic neurons in the intermediolateral horn (preganglionic cells)
  C. Internuncial cells of the spinal cord
  D. Gamma motoneurons
  E. Cells in the posterior (dorsal) horn   Ref. p. 648

746. OF THE SENSORY FIBERS ENTERING THE SPINAL CORD THROUGH THE DORSAL ROOTS AND MAKING SYNAPTIC CONTACT WITH NEURONS LOCATED AT APPROXIMATELY THE SAME SEGMENTAL LEVEL, THE MAJORITY SYNAPSE DIRECTLY WITH:
  A. Anterior motoneurons
  B. Sympathetic neurons
  C. Internuncial cells
  D. Gamma motoneurons
  E. Cells in the posterior (dorsal) horn   Ref. p. 648

747. THE WITHDRAWAL REFLEX:
  A. Is monosynaptic
  B. Is polysynaptic
  C. Is more rapid than the stretch reflex
  D. Is antigravitational
  E. Does not appear if the spinal cord is separated from suprasegmental levels   Ref. p. 648

748. GAMMA MOTOR NERVE FIBERS INDUCE THE:
  A. Contraction of the end portions of the intrafusal fibers in the muscle spindle
  B. Contraction of the mid-portions of the intrafusal fibers -where nuclei are most abundant-in the muscle spindle
  C. Contraction of the extrafusal skeletal muscle fibers
  D. Relaxation of the tone of the intrafusal fibers
      Ref. p. 649

749. THE MAIN FUNCTION OF THE PRIMARY RECEPTORS (ANNULOSPIRAL) IN THE MUSCLE SPINDLE CONSISTS OF:
  A. Responding to very marked muscle tensions which then results in inhibition of the contraction
  B. Responding to total tension imposed on the muscle but not on the rate of change of length of the receptor
  C. Responding to exquisitely small changes in tension and to the rate of change of elongation of the receptors as well
  D. Transmitting the sensations of pain in muscle to the central nervous system   Ref. p. 649

SECTION IV - PHYSIOLOGY OF THE NERVOUS SYSTEM

750. THE SECONDARY RECEPTOR (FLOWER SPRAY) OF THE MUSCLE SPINDLE:
    A. Conveys to the spinal cord action potentials only related to the extent of elongation of the spindle without correlation with the rate of change of length
    B. Sends impulses to the spinal cord in fibers of the widest diameter found for afferent fibers
    C. Is the receptor that responds with the least interval after stimulation of all muscle receptors
    D. Originates signals related to changes occurring in tendons but not in the muscle   Ref. p. 649

751. ACTIVATION OF GAMMA MOTOR NEURONS CAUSES:
    A. The muscle spindles to be relaxed from a previous tonic state of contraction
    B. The muscle spindles to be shortened
    C. Inhibition of activity of the annulospiral receptors
    D. Inhibition of the activity of the flower spray receptors
                                                Ref. p. 649

752. IN ADDITION TO THE MOTONEURONS IN THE SPINAL CORD, THE MAJOR STRUCTURE IN THE CENTRAL NERVOUS SYSTEM WHICH RECEIVES INPUT FROM THE ANNULOSPIRAL RECEPTORS OF MUSCLES IS THE:
    A. Thalamus   D. Putamen
    B. Cerebellum   E. Temporal lobe
    C. Caudate nucleus   Ref. p. 650

753. THE MAJOR CHANGE OCCURRING IN SKELETAL MUSCLE "RECORDED" BY THE GOLGI TENDON ORGAN IS THE INCREASE IN MUSCLE:
    A. Thickness   D. Tension
    B. Oxygen supply   E. Length
    C. ATP stores   Ref. p. 650

754. IMPULSES ARISING FROM RECEPTORS FOUND IN TENDONS (GOLGI TENDON ORGANS) ARRIVE AT THE SPINAL CORD:
    A. Through sympathetic afferents
    B. Fibers that transmit pain sensations
    C. Through fibers that transmit temperature sensations
    D. Through large type A alpha nerve fibers similar to those that transmit impulses from the muscle spindles
    E. At a very slow rate of conduction   Ref. p. 650

755. THE RESPONSE OBTAINED IN A MYOTATIC REFLEX CONSISTS OF THE:
    A. Shortening of the muscle previously elongated
    B. Relaxation of the muscle previously contracted
    C. Contraction of the antagonist muscle when the agonist muscle is elongated
    D. Extension of the contralateral limb when the ipsilateral limb is subjected to a nociceptive stimulus   Ref. p. 650

SECTION IV - PHYSIOLOGY OF THE NERVOUS SYSTEM

756. THE BRIEF SUDDEN ELONGATION OF THE EXTENSOR MUSCLES OF THE THIGH CAUSED BY TAPPING THE PATELLAR TENDON IS AN EXAMPLE OF A:
   A. Static stretch reflex
   B. Dynamic stretch reflex
   C. Nociceptive reflex
   D. Negative stretch reflex   Ref. pp. 651, 653

757. THE GAMMA MOTOR NEURON (EFFERENT) ACTIVITY LEADS TO:
   A. Shortening of the extrafusal fibers to the same length as the intrafusal fibers and no more
   B. An excessive shortening of extrafusal fibers which overshoots the desired length
   C. Relaxation of extrafusal fibers
   D. Relaxation of intrafusal fibers   Ref. p. 652

758. MUSCLE TONE IN ANTIGRAVITATIONAL MUSCLES:
   A. Is due to a continuous basal rate of firing of alpha motor neurons to extrafusal fibers
   B. Is basically independent of the rate of discharge of gamma motor neurons
   C. Should be preserved intact if afferent impulses arriving to the spinal cord from annulospiral receptors is discontinued (section of dorsal roots)
   D. Depends entirely on voluntary control of muscle length
             Ref. p. 652

759. IT IS LOGICAL TO SPECULATE THAT DURING A DELICATE AND PRECISE MOVEMENT OF AN UPPER EXTREMITY:
   A. The gamma motor fibers are totally suppressed in their activity leaving only alpha motor neurons to conduct the sequence of muscle contractions
   B. The gamma motor neurons increase their activity and the tone of the muscles about to undertake the delicate motions
   C. Alpha motor neuron activity is greatly suppressed in favor of automatic reflexes
   D. Golgi tendon organs are mainly responsible for maintaining the afferent input to the cord   Ref. p. 653

760. AN INCREASED ACTIVITY OF MUSCLE STRETCH REFLEXES:
   A. May be associated with clonus
   B. Will disappear if the facilitatory areas in the brain stem are disconnected from suprasegmental levels (cerebral cortex, cerebellum)
   C. Is not associated with spasticity
   D. Will not disappear if alpha motor neuron functional activity is inhibited
             Ref. p. 654

761. IN THE SO-CALLED LENGTHENING REACTION:
   A. The contracting muscle fibers are suddenly relaxed through inhibition of alpha motor neurons
   B. The muscle is contracted further as the tendon is elongated
   C. Increased activity of Golgi tendon organs (receptors) stimulate internuncial cells in the cord and stimulate further the alpha motor neurons
   D. There is lengthening of muscle and tendon in response to a suprasegmental inhibition of gamma motor neurons
             Ref. p. 654

SECTION IV - PHYSIOLOGY OF THE NERVOUS SYSTEM

762. A FLEXOR (NOCICEPTIVE) REFLEX IS KNOWN TO BE:
    A. Monosynaptic
    B. Slightly delayed by internuncial cells in the spinal cord
    C. Suppressed if suprasegmental (above the medulla oblongata) centers are separated from the cord
    D. Antigravitational
    E. Insensitive to light stimuli (touch)   Ref. p. 655

763. IT IS KNOWN THAT IN A FLEXOR REFLEX ELICITED BY A STRONG NOXIOUS STIMULUS:
    A. No fatigue occurs in the response
    B. The afterdischarge is very brief (1-2 milliseconds)
    C. A single synapse intervenes between afferent stimuli and alpha motor neuron
    D. The antagonist of the flexing muscle in the same limb is stimulated to induce a compensatory extension
    E. The afterdischarge is characteristically prolonged
    Ref. p. 655

764. AFTERDISCHARGE IS KNOWN TO BE PROLONGED TO THE GREATEST EXTENT IN THE CASE OF THE:
    A. Myotatic reflex   D. Lengthening reaction
    B. Flexor withdrawal reflex   E. Negative stretch reflex
    C. Crossed extensor reflex   Ref. p. 657

765. THE PHENOMENON KNOWN AS RECIPROCAL INHIBITION:
    A. May suppress a nociceptive reflex on one extremity if the contralateral limb is stimulated to an even greater degree by another nociceptive stimulus
    B. Inhibits the agonist muscles only
    C. Causes strong inhibition of the muscles after a contraction-due to a strong nociceptive stimulus-is over
    D. Is mediated by monosynaptic reflexes
    Ref. Fig. 56-9, p. 657

766. IN WHAT IS KNOWN AS POSITIVE SUPPORTIVE REACTION:
    A. The limbs may support the whole body if subjected to nociceptive stimuli
    B. The portions of the central nervous system rostral to the spinal cord such as the cerebral cortex, cerebellum, play an essential role
    C. A gentle tactile stimulus to the distal portion of the extremity causes the extremity to be flexed
    D. A tactile stimulus to the foot pad of a decerebrate animal will cause the whole limb to be extended   Ref. p. 658

767. IN THE RECOVERY PERIOD THAT FOLLOWS AN EXPERIMENTAL TRANSECTION OF THE CERVICAL CORD OF A DOG, THE SPONTANEOUS MOVEMENTS OF THE LIMBS OF THE ANIMAL WHICH TEND TO PLACE IT STANDING ON ITS FEET (RIGHTING REFLEXES), EVEN THOUGH IT MAY NOT SUCCEED COMPLETELY, CAN BE EXPLAINED AS DUE TO:
    A. An incomplete transection of the descending corticospinal tracts
    B. Simple monosynaptic reflexes
    C. A series of myotatic reflexes
    D. The complex integrative reflex function that the spinal cord has
    Ref. p. 658

SECTION IV - PHYSIOLOGY OF THE NERVOUS SYSTEM

768. THE REPETITIVE TO AND FRO MOVEMENTS OF A LIMB THAT ARE SEEN IN THE SO-CALLED SCRATCH REFLEX (WHICH MAY BE FOUND IN ANIMALS WHOSE SPINAL CORD HAS BEEN TRANSECTED IN THE CERVICAL REGION):
    A. Depends on intact afferent signals from the skin
    B. Depends on myotatic reflexes only
    C. Depends on withdrawal reflexes only
    D. May occur even if dorsal roots are transected, once the movements have started   Ref. p. 659

769. A MUSCLE CRAMP THAT DOES NOT SUBSIDE AFTER SEVERAL HOURS MAY BE RELIEVED BY VOLUNTARILY:
    A. Contracting further the spastic muscle
    B. Relaxing voluntarily the spastic muscle
    C. Hyperventilating
    D. Contracting isometrically the muscles of the opposite limb
    Ref. p. 660

770. THE SO-CALLED MASS REFLEX:
    A. Does not occur in an animal whose spinal cord has been separated from suprasegmental levels
    B. Has the pattern of a massive myotatic reflex and leads to an extensor rigidity
    C. Has the pattern of a massive flexor reflex with emptying of rectum and urinary bladder
    D. Is usually elicited by a mild tactile stimulus similar to that which elicits the scratch reflex   Ref. p. 660

771. THE REASON WHY THE SPINAL CORD HAS, FOR A PERIOD OF TIME, LITTLE OR NO REFLEX ACTIVITY IMMEDIATELY FOLLOWING ITS ACUTE COMPLETE TRANSECTION, IS DUE TO THE FACT THAT:
    A. Spinal motoneurons are dependent on suprasegmental stimuli to a very large extent, and their removal leads to transient hypofunction of the motor neurons
    B. The vascular supply to the spinal cord is always interrupted by the transection
    C. Afferent signals from receptors are inhibited by the transection
    D. The reflex activity of the transected cord is practically zero
    Ref. p. 660

772. THE FACILITATORY PORTION OF THE RETICULAR FORMATION:
    A. Is restricted to the pons and cerebral peduncles; it does not extend to the thalamus or to the medulla oblongata
    B. Initiates impulses only when stimulated by suprasegmental levels (cerebral cortex, cerebellum)
    C. Is inhibited by suprasegmental levels (cerebral cortex, cerebellum)
    D. Facilitates the flexor withdrawal reflex action of the spinal cord
    Ref. p. 663

SECTION IV - PHYSIOLOGY OF THE NERVOUS SYSTEM

773. IN THE NORMAL ANIMAL, INHIBITION OF THE FACILITATORY PORTION OF THE RETICULAR FORMATION:
    A. Is continuously present at all times at a basal level
    B. Occurs only during sleep
    C. Occurs only if the animal receives an anesthetic dose of a sedative
    D. Becomes fully active only if the cerebellar connections to the brain stem are interrupted
    E. Becomes fully active only if the brain stem is transected at the level of the superior colliculi leaving the cerebellum attached to the pons
    Ref. p. 663

774. A METHODICAL TRANSVERSE SECTIONING OF THE BRAIN STEM OF A CAT, STARTING AT THE CEREBRAL PEDUNCLES, AND PROGRESSING IN A CAUDAL DIRECTION, WILL TRANSFORM AN INITIAL DECEREBRATE RIGIDITY INTO A FLACCID PARALYSIS. THIS NEW FLACCIDITY IS DUE TO THE REMOVAL OF MOST OF THE FACILITATORY AREA OF THE RETICULAR FORMATION, WHICH BECOMES "DISCONNECTED" FROM THE SPINAL CORD. THE BRAIN STEM NUCLEI BELOW WHICH A TRANSECTION CAUSES FLACCIDITY TO APPEAR IN A DECEREBRATE RIGID CAT IS THE:
    A. Nucleus of the phrenic nerve
    B. Vestibular nuclei
    C. Nucleus of the VII cranial nerve (facial)
    D. Nucleus of the III cranial nerve (oculomotor)
    E. Nucleus of Edinger-Westphal   Ref. p. 663

775. ELECTRICAL STIMULATION OF THE VENTROMEDIAL PORTION OF THE MEDULLA IN THE AREA OF THE MEDULLARY RETICULAR FORMATION CAUSES:
    A. Further increase of rigidity in a decerebrate cat
    B. No change in muscle tone from a normal basal level
    C. Active motion of the hindlimbs
    D. Active motions of the forelimbs
    E. Flaccidity in a rigid decerebrate cat   Ref. p. 663

776. IN HUMANS, IN ORDER TO PLACE THE EXTERNAL SEMICIRCULAR CANALS ON A HORIZONTAL PLANE, THE SUBJECT SITTING IN A CHAIR WITH A STRAIGHT BACK PERPENDICULAR TO THE SEAT MUST HAVE HIS HEAD:
    A. Tilted backward 60^0 from the normal vertical position
    B. Tilted forward 60^0 from the normal vertical position
    C. In a normal vertical position
    D. Tilted forward 30^0 from the normal vertical position
    E. Tilted backward 30^0 from the normal vertical position
    Ref. p. 664

777. THE ANATOMICAL DISPOSITION OF THE RIGHT ANTERIOR SEMICIRCULAR CANAL IS ARRANGED IN SUCH A MANNER THAT IT IS:
    A. Parallel to the left anterior semicircular canal
    B. Perpendicular to the left posterior semicircular canal
    C. With its axis directed parallel to the right external auditory canal
    D. Parallel to the left posterior semicircular canal
    E. Parallel to the right posterior semicircular canal
    Ref. p. 664

SECTION IV - PHYSIOLOGY OF THE NERVOUS SYSTEM

778. RECEPTOR CELLS LOCATED IN THE UTRICLE HAVE HAIR CELLS WHOSE CILIA ARE:
    A. Embedded in what is known as the crista ampullaris
    B. Embedded in a gelatinous mass known as cupula
    C. Stimulated by the weight of otoconia
    D. Specifically stimulated by moving endolymphatic fluid which displaces them laterally
    E. Sensitive to changes in acceleration but not to weight of otoconia
    Ref. p. 664

779. THE PHYSICAL CHANGE WHICH ORIGINATES IMPULSES IN THE AMPULLAR CRISTAE OF SEMICIRCULAR CANALS IS THE:
    A. Increased temperature of endolymphatic fluid
    B. Decreased pressure of endolymphatic fluid
    C. Increased pressure of endolymphatic fluid
    D. Inertia of endolymphatic fluid as it lags behind lateral movements of the osseous canals   Ref. p. 664

780. BENDING OF EACH KINOCILIUM:
    A. Causes stimulation regardless of the direction of the bending
    B. Causes inhibition regardless of the direction of the bending
    C. Stimulates the hair cell when the bending is in one direction and inhibits the same cell when the bending occurs in the opposite direction
    D. Takes place in one direction only, for each kinocilium is rigid for any other direction   Ref. p. 664

781. TO PRESERVE A NORMAL CONTROL OF RAPID MOVEMENTS OF THE EXTREMITIES THERE MUST BE AN INTACT LINKAGE BETWEEN VESTIBULAR NUCLEI AND THE:
    A. Flocculonodular lobe of the cerebellum
    B. Nucleus of the III cranial nerve
    C. Nucleus of the VII cranial nerve
    D. Nucleus of the trigeminal nerve
    E. Medial longitudinal fasciculus   Ref. p. 665

782. A SUDDEN FORWARD ACCELERATION OF THE HEAD CREATES IN THE CENTRAL NERVOUS SYSTEM THE SENSATION THAT THE WHOLE BODY IS:
    A. Falling forward
    B. Falling backward
    C. Rotating along its longest axis
    D. In steady motion forward
    E. In steady motion to the side of the body where gravity is applying its force   Ref. p. 665

783. THE PORTION OF THE INNER EAR MOST SENSITIVE TO <u>LINEAR</u> ACCELERATION (FORWARD, BACKWARD AND TO THE LATERAL SIDE) IS THE:
    A. Sacule
    B. Anterior semicircular canal
    C. Lateral semicircular canal
    D. Posterior semicircular canal
    E. Utricle   Ref. p. 665

SECTION IV - PHYSIOLOGY OF THE NERVOUS SYSTEM   151

784. AFTER ABOUT 20 SECONDS OF ROTATING (ONE REVOLUTION EVERY FOUR SECONDS ON A ROTATING CHAIR) AT A UNIFORM SPEED A BLINDFOLDED NORMAL PERSON WHOSE HEAD IS KEPT UPRIGHT, THE PERSON WILL MOST LIKELY REPORT THAT HE FEELS AS IF HE WERE:
   A. Falling backward
   B. Falling forward
   C. Not moving at all
   D. Falling to the side toward which is is being moved
   Ref. p. 666

785. ASSUME THAT THE HAIR CELLS OF A GIVEN SEMICIRCULAR CANAL INCREASE THEIR RATE OF FIRING WHEN THE CIRCULAR MOVEMENTS OF ENDOLYMPHATIC FLUID ARE INITIATED. WHEN THE MOVEMENTS STOP ABRUPTLY, THE HAIR CELLS RESPOND BY:
   A. Increasing their rate even further
   B. Maintaining their previous rate
   C. Dropping the rate transiently to zero
   D. Decreasing the rate minimally from the rate found when acceleration was began   Ref. Fig. 57-5, p. 666

786. THE LOSS OF FUNCTION OF THE SEMICIRCULAR CANALS WILL IMPAIR TO A GREATEST EXTENT THE CAPACITY TO MAINTAIN EQUILIBRIUM:
   A. In rapid body movements
   B. In slow body movements
   C. When the body is at rest in the supine position
   D. When the body is at rest in the prone position
   Ref. p. 666

787. THE EXPERIMENTAL ABLATION OF THE FLOCCULONODULAR LOBES OF THE CEREBELLUM WILL CAUSE THE LEAST IMPAIRMENT OF FUNCTION OF RECEPTORS LOCATED IN THE:
   A. Macula (utricle)
   B. Ampullae of the horizontal semicircular canals
   C. Ampullae of the anterior semicircular canals
   D. Ampullae of the posterior semicircular canals
   Ref. p. 667

788. THE IMMINENT FALL TO THE RIGHT DUE TO EXTRANEOUS FORCES OR TO SUDDEN LOSS OF EQUILIBRIUM CAUSES THE REFLEX:
   A. Flexion of the right lower extremity
   B. Flexion of the left lower extremity
   C. Hyperextension of the head
   D. Flexion of the head
   E. Extension of the right lower extremity
   Ref. p. 667

789. A SUDDEN ACCELERATION OF THE HEAD TOWARD THE GROUND WILL RESULT IN THE REFLEX:
   A. Hyperextension of the forelimbs
   B. Flexion of the forelimbs
   C. Flexion of the hindlimbs
   D. Flexion of the whole vertebral column
   Ref. p. 667

SECTION IV - PHYSIOLOGY OF THE NERVOUS SYSTEM

790. WHEN A PERSON WHO IS SITTING ON A ROTATING CHAIR IS SUDDENLY ACCELERATED TO ROTATE TO THE RIGHT SIDE WITH HIS HEAD TILTED 30° FORWARD, HIS LEFT AND RIGHT EYES WILL:
    A. Move slowly to the left of the person's visual field
    B. Move slowly to the right of the person's visual field
    C. Move rapidly to the left of the person's visual field
    D. Not move at all until the velocity has become a steady one
    Ref. p. 667

791. IN A BARANY TEST, THE SUBJECTIVE FEELING PRESENT IMMEDIATELY AFTER THE MOVEMENT HAS STOPPED MAKES THE PERSON BELIEVE THAT:
    A. He is still rotating in the same direction as before stopping
    B. He is still rotating in the opposite direction as before stopping
    C. His body is absolutely quiescent
    D. His body is subjected to a to-and-fro frontal motion
    Ref. p. 668

792. IF THE INTRODUCTION OF ICE WATER IN THE RIGHT EXTERNAL EAR CANAL FAILS TO CAUSE NYSTAGMUS, WHILE THE SAME PROCEDURE ON THE LEFT SIDE CAUSES NYSTAGMUS TO APPEAR, IT PROBABLY MEANS THAT THE:
    A. Utricle has ceased to function on the right side
    B. Use of warm water will probably succeed in revealing nystagmus on the right side
    C. Endolymph in the right semicircular canals has a greater density than that on the left side
    D. Vestibular pathways or the internal ear are damaged by organic disease
    Ref. p. 668

793. THE SENSE OF POSITION OF THE HEAD WITH RESPECT TO THE WHOLE BODY IS MAINTAINED BY RECEPTORS LOCATED IN THE JOINTS AND MUSCLES OF THE NECK. THIS INFORMATION IS FUNNELED INTO THE CENTRAL NERVOUS SYSTEM WHERE COMPLEX REFLEXES OPERATE A COMPENSATORY RESPONSE TO THE ORIGINAL CHANGES. IT IS TRUE THAT NECK REFLEXES:
    A. Will cause hyperextension of the forelimbs when the head tends to drop in the otherwise intact animal
    B. Will cause hyperextension of the forelimbs when the head tends to drop if the vestibular nuclei are destroyed
    C. Are synergistic with the vestibular reflexes
    D. Will not manifest themselves unless the vestibular nuclei are destroyed as the vestibular reflexes are antagonistic to neck reflexes
    Ref. p. 668

794. MAINTENANCE OF EQUILIBRIUM DEPENDS ON INFORMATION REACHING THE CENTRAL NERVOUS SYSTEM FROM VARIOUS SOURCES LISTED BELOW, EXCEPT:
    A. The retina
    B. The cochlear nerve
    C. The vestibular apparatus
    D. The proprioceptive receptors in muscles and joints
    E. Exteroceptive receptors in the skin   Ref. p. 668

SECTION IV - PHYSIOLOGY OF THE NERVOUS SYSTEM

795. IT IS BELIEVED TRUE THAT ASIDE FROM CEREBELLAR AND BRAIN STEM INTEGRATION OF THE STATE OF EQUILIBRIUM, THERE IS AN AREA IN THE CEREBRUM WHERE THE CONSCIOUS FEELING OF EQUILIBRIUM OR LACK OF EQUILIBRIUM IS LOCALIZED. THIS AREA IS BELIEVED TO BE LOCATED IN THE:
    A. Hypothalamus
    B. Corpus callosum
    C. Premotor cortex
    D. Occipital cortex (areas 18 and 19)
    E. Temporal lobe (next to the area of hearing)
    Ref. p. 669

796. FROM STIMULATION STUDIES IT IS PROBABLY TRUE THAT THE MAGNOCELLULAR PORTION OF THE RED NUCLEUS IS IN SOME MANNER CONCERNED WITH MOVEMENTS OF:
    A. A very fine nature in the distal portions of the extremities
    B. The eyes laterally, upward and downward
    C. The small intestine
    D. The head and upper trunk backward
    E. The large intestine
    Ref. p. 669

797. THE LOSS OF FUNCTION OF THE SUBSTANTIA NIGRA SEEMS TO DECREASE SPECIFICALLY THE:
    A. Physiological sensitivity of the corpus callosum
    B. Control of the hypothalamus on the autonomic nervous system
    C. Control of the gamma activating system of the muscle spindles
    D. Function of the alpha motor neurons
    E. Control of the corticospinal tract on alpha motor neurons
    Ref. p. 670

798. A CAT WHOSE NEURAXIS IS TRANSECTED BETWEEN THE THALAMUS AND SUBTHALAMIC NUCLEI WILL NOT BE ABLE TO:
    A. Stand
    B. Walk
    C. Overcome an obstacle
    D. Retain complex spinal reflexes
    E. Stand up and right itself
    Ref. p. 670

799. DECORTICATION IN A CAT CAUSES MOTOR DISTURBANCES THAT HELP TO UNDERSTAND THE MOTOR FUNCTIONS OF THE BASAL GANGLIA. A DECORTICATE CAT IS ABLE TO PERFORM ALL THE FOLLOWING MOVEMENTS, EXCEPT:
    A. Walking
    B. Eating
    C. Movements associated with mating
    D. Discrete movements of the extremities
    E. Flexion and extension of the head
    Ref. p. 671

SECTION IV - PHYSIOLOGY OF THE NERVOUS SYSTEM

800. IF A DECORTICATE ANIMAL IS SUBSEQUENTLY SUBJECTED TO THE REMOVAL OF A PORTION OF ITS BASAL GANGLIA, THE ONLY COMPLEX MOVEMENTS STILL REMAINING ARE THOSE THAT ALLOW THE ANIMAL TO:
    A. Mate
    B. Eat
    C. Move the extremities with fine purposeful movements
    D. Walk, without being able to overcome obstacles
    E. Fight
    Ref. p. 671

801. A HUMAN WHOSE CEREBRAL CORTEX IS SELECTIVELY DAMAGED WITHOUT PATHOLOGICAL INVOLVEMENT OF THE BASAL GANGLIA IS STILL ABLE TO:
    A. Think
    B. Recall
    C. Perform arithmetical calculations
    D. Maintain equilibrium
    E. Perform fine movements with his hands
    Ref. p. 671

802. A CONSEQUENCE OF THE WIDESPREAD DAMAGE OF THE BASAL GANGLIA IN HUMANS IS A(N):
    A. Increase in facilitatory impulses from the reticular formation of the brain stem to the myotatic reflex in the spinal cord
    B. Flaccid paralysis
    C. Increased capacity to perform fine rapid movements
    D. Decreased visual acuity due to damage of part of the visual pathways
    Ref. p. 671

803. THE MOST IMPORTANT EFFERENTS OF THE STRIATE BODY (CAUDATE NUCLEUS AND PUTAMEN) REACH AND ESTABLISH SYNAPTIC CONNECTIONS WITH THE:
    A. Motor cortex
    B. Visual cortex
    C. Auditory cortex
    D. Rhinencephalic cortex
    E. Pituitary
    Ref. p. 671

804. THROUGH ITS CORTICAL CONNECTIONS AN INFLUENCE OF THE STRIATE BODY IS THAT OF:
    A. Increasing visual function
    B. Increasing auditory function
    C. Increasing hypothalamic releasing factor secretion
    D. Inhibiting coarse unconscious intentional movements
    Ref. p. 671

805. A FINAL EFFERENT PATHWAY FOR THE INFLUENCE OF THE GLOBUS PALLIDUS IS THE:
    A. Olivocerebellar tract
    B. Restiform body
    C. Medial lemniscus
    D. Optic radiations
    E. Reticulospinal tract
    Ref. p. 671

806. SLOW ROTATING MOVEMENTS OF ONE OR MORE OF THE EXTREMITIES WITH HYPEREXTENSION OF FINGERS IS CHARACTERISTIC OF THE MOTOR DYSFUNCTION SEEN IN A DISEASE KNOWN AS:
    A. Chorea
    B. Hemiballism
    C. Parkinson's disease
    D. Catatonia
    E. Athetosis
    Ref. p. 672

SECTION IV - PHYSIOLOGY OF THE NERVOUS SYSTEM   155

807. THE AREA OF THE CENTRAL NERVOUS SYSTEM WHICH IS CRITICALLY DAMAGED IN A CASE OF UNILATERAL HEMIBALLISMUS IS THE:
A. Substantia nigra
B. Restiform body
C. Superior olive
D. Subthalamic nucleus
E. Dorsomedial nucleus of the hypothalamus
Ref. p. 672

808. BECAUSE THE CHARACTERISTIC LESION IN PARKINSON'S DISEASE DESTROYS PARTS OF THE SUBSTANTIA NIGRA, A MAJOR DYSFUNCTION CONSISTS OF THE:
A. Excessive excitation of the gamma efferent system
B. Inactivation of the gamma efferent system
C. Inactivation of the alpha motor neurons
D. Inactivation of the cerebellar control of movements
E. Inactivation of afferent signals through the medial lemniscus
Ref. p. 672

809. THE RIGIDITY SEEN IN PARKINSON'S DISEASE IS ASSOCIATED WITH:
A. Excessive "bombardment" over alpha motor neurons
B. Hypofunction of the cerebellum
C. Hypofunction of the motor cortex
D. Hypofunction of proprioceptive ascending fibers to the cerebellum
Ref. p. 672

810. RIGIDITY OF THE PATIENT WITH PARKINSON'S DISEASE IS:
A. Similar to that of the decerebrate animal
B. Exerted only in protagonist muscles with inhibition of the antagonist muscles
C. Associated with hyperactive myotatic reflexes
D. Exerted on protagonist and antagonist muscles as well
Ref. p. 672

811. RELIEF OF THE MUSCLE RIGIDITY FOUND IN PARKINSON'S DISEASE MAY BE SURGICALLY OBTAINED BY DESTROYING BY ELECTROCOAGULATION THE:
A. Red nucleus
B. Ventrolateral nucleus of the hypothalamus
C. Ventrolateral nucleus of the thalamus
D. Motor cortex of the dominant side
E. Motor cortex of the non-dominant side
Ref. p. 673

812. THE PREVIOUS NEUROSURGICAL PROCEDURE (ELECTROCOAGULATION OF A LOCALIZED AREA WITHIN THE BRAIN) PRESUMABLY BLOCKS THE OUTPUT, WHICH, IN THE FORM OF AN EXCESSIVE FEEDBACK, ORIGINATES IN THE:
A. Basal ganglia
B. Motor cortex
C. Premotor cortex
D. Medullary portion of the reticular formation
E. Substantia nigra
Ref. p. 673

SECTION IV - PHYSIOLOGY OF THE NERVOUS SYSTEM

813. THE SUCCESSFUL ADMINISTRATION OF L-DOPA TO PATIENTS WITH PARKINSON'S DISEASE MAY REPLENISH NEURONAL STORES OF:
    A. Acetylcholine
    B. Epinephrine
    C. Dopamine
    D. 5-Hydroxytryptamine
    E. GABA (gamma-amino-butyric acid)   Ref. p. 673

814. THE SO-CALLED "PLACING REACTION" AND "HOPPING REACTION" ARE PRESENT IN AN ANIMAL WHO AFTER BEING DEPRIVED OF ITS VISUAL FUNCTION STILL HAS AN INTACT:
    A. Flocculonodular lobe of the cerebellum
    B. Dorsomedial nucleus of the hypothalamus
    C. Ventromedial nucleus of the hypothalamus
    D. Cingulate gyrus
    E. Cerebral cortex   Ref. p. 674

815. ALL OF THE FOLLOWING REFLEXES ARE TO BE FOUND PRESERVED IN A SPINAL ANIMAL, EXCEPT:
    A. Myotatic reflex
    B. Flexor reflex
    C. Scratch reflex
    D. Optokinetic reflex
    E. Neck reflexes
    Ref. pp. 667, 674

816. MOVEMENT OF AN EXTREMITY MAY RESULT FROM A MINIMAL ELECTRICAL STIMULUS APPLIED TO THE CEREBRAL CORTEX IF THE STIMULATION IS LOCALIZED:
    A. Where afferent radiations from the thalamus are most abundant
    B. In the anterior pole of the frontal lobe
    C. At the insula (of Reil)
    D. Where Betz cells are present
    E. In the premotor areas (6 and 8)   Ref. p. 677

817. THE SO-CALLED "PYRAMIDAL" TRACT CONTAINS AXONS WHICH ORIGINATE:
    A. Solely in the area where Betz cells are located
    B. Solely in the motor and premotor areas of the frontal lobe
    C. In part in the somesthetic cortex (parietal lobe) as well as in the premotor and motor areas
    D. In all of the cerebral cortex, including the frontal, parietal, temporal and occipital lobes   Ref. p. 677

818. IF A COUNT IS MADE OF ALL AXONS WHICH DESCEND FROM THE BRAIN AND PASS THROUGH THE MEDULLARY PYRAMIDS (IN THE BRAIN STEM) IN THEIR PATH TO THE SPINAL CORD, THE FIBERS THAT CAN BE TRACED BACK TO THE BETZ CELLS AMOUNT TO:
    A. 3%
    B. 30%
    C. 50%
    D. 75%
    E. 100%
    Ref. p. 678

819. THE BULK OF THE AXONS IN THE PYRAMIDAL TRACT HAVE SYNAPTIC CONTACT WITH:
    A. Motor neurons in the anterior horns of the spinal cord
    B. Cells in the intermedio-lateral column of the spinal cord
    C. Cells located in the posterior horn of the spinal cord
    D. Internuncial cells in the spinal cord
    E. Cells in the brain stem   Ref. p. 678

SECTION IV - PHYSIOLOGY OF THE NERVOUS SYSTEM 157

820. A LARGE FRACTION OF COLLATERALS OF THE CORTICOSPINAL TRACT HAVE DIRECT SYNAPTIC CONTACT WITH THE:
A. Occipital lobe
B. Hypothalamus
C. Cerebellum
D. Temporal lobe
E. Insula
Ref. p. 678

821. A SELECTIVE CONTRACTION OF MUSCLES IN THE RIGHT FOOT WILL RESULT FROM THE ELECTRICAL STIMULATION OF THE AREA OF THE MOTOR CORTEX WHICH LIES IN THE LEFT HEMISPHERE:
A. In the lower third of the anterior central gyrus (lateral side)
B. In the middle third of the anterior central gyrus (lateral side)
C. In the upper third of the anterior central gyrus (lateral side)
D. On the medial side of the anterior central gyrus between the dorsal margin and the corpus callosum   Ref. p. 679

822. THE VOCAL CORDS HAVE AN AREA OF CORTICAL REPRESENTATION WHICH IS APPROXIMATELY SIMILAR IN SIZE TO THAT OF THE:
A. Tongue
B. Eyelids
C. Neck
D. Knee
E. Elbow
Ref. p. 679

823. OF THE FOLLOWING, THE MUSCLES OF THE UPPER EXTREMITY WHICH ARE REPRESENTED THROUGH A LARGER AREA OF THE MOTOR CORTEX ARE THOSE THAT:
A. Extend the shoulder
B. Flex the humerus
C. Extend the forearm
D. Flex the fingers   Ref. p. 679

824. THE ELECTRICAL STIMULATION OF A GIVEN POINT IN THE CEREBRAL CORTEX WHICH RESULTS IN THE CONTRACTION OF MUSCLES OF ONE LIMB WILL:
A. Cause the same force of contraction regardless of the strength of the stimulus
B. Cause the extremity to adopt a final position which depends on the initial position of the limb before the stimulation
C. Cause a tetanic contraction of that limb
D. Make the limb adopt a final position which depends only on the exact location of the stimulation but not on the original position adopted by the limb prior to stimulation   Ref. p. 679

825. ALL OF THE FOLLOWING CHARACTERIZE THE PHYSIOLOGICAL FEATURES OF MOTOR FUNCTION OF THE SO-CALLED "PREMOTOR AREA" WITH THE EXCEPTION OF ONE. THUS IT WOULD BE IN ERROR TO SAY THAT THE PREMOTOR AREA IS CHARACTERIZED BY:
A. Responses (motor) which are slow to develop in response to stimulation
B. Responses which require larger stimuli than those that elicit a response in the "motor area"
C. Responses which are more discrete than those obtained by stimulating the "motor area"
D. Responses which require a more prolonged stimulation of an effective stimulus
E. Responses which are more complex than those obtained by stimulating the "motor area"   Ref. p. 680

SECTION IV - PHYSIOLOGY OF THE NERVOUS SYSTEM

826. EXPERIMENTAL LESIONS OF THE MOTOR CORTEX ARE SOMETIMES ASSOCIATED WITH SPASTICITY AND SOMETIMES WITH FLACCIDITY. FLACCIDITY OF MUSCLES IS MOST LIKELY TO RESULT FROM LESIONS SPECIFICALLY RESTRICTED TO THE:
    A. Posterior part of the area pyramidalis
    B. Anterior part of the area pyramidalis
    C. Premotor area
    D. Sensory area   Ref. p. 681

827. A LOCALIZED DESTRUCTION OF THE BRAIN STEM PYRAMID ON THE RIGHT SIDE CAUSES:
    A. An ipsilateral paralysis of all coarse movements but preservation of fine movements on the right side
    B. Contralateral loss of fine movements
    C. Loss of sensation of vibration on the contralateral side
    D. The interruption of the vestibulospinal tract on the same side
    E. The interruption of the bulk of the motor function of the extremities on the same side of the lesion   Ref. p. 681

828. AFTER A LESION WHICH INTERRUPTS THE CORTICOSPINAL TRACT AT THE LEVEL OF THE INTERNAL CAPSULE, THE MOTOR FUNCTION WHICH IS BEST PRESERVED IS THAT WHICH SERVES TO:
    A. Support the whole body against gravity
    B. Change rapidly the direction of movement of the whole body
    C. Execute fine movements with the fingers of the hand of the same side
    D. Walk   Ref. p. 681

829. ALL OF THE FOLLOWING REPRESENT SOURCES OF AFFERENT PATHWAYS TO THE CEREBELLUM, EXCEPT:
    A. Inferior olive   D. Spinal cord
    B. Vestibular nuclei   E. Dentate nucleus
    C. Reticular formation of the pons   Ref. p. 683

830. ALL OF THE FOLLOWING STATEMENTS CONCERNING THE CEREBELLUM ARE CORRECT, EXCEPT:
    A. Voluntary movements do not originate in the cerebellum but are only modified by this organ
    B. The cerebellum is stimulated by the motor activity which originates in the cerebral cortex
    C. A major portion of the neuronal activity leaving the cerebellum ends up modifying the cerebral motor cortex activity
    D. Much of the receptor activity of the spindles and Golgi tendon organs reaches the cerebellum through the spinocerebellar tracts
    E. The cerebellum has no effect on the reticular formation influence on the myotatic reflex of the spinal cord
                                   Ref. p. 685

831. IN THE EXECUTION OF A MOVEMENT WHICH BRINGS THE EXTENDED ARM AND INDEX FINGERS POINTING AND AIMING TOWARD A GIVEN LINE, AT WHICH THE FINGER HAS TO STOP, THE CEREBELLUM IS RESPONSIBLE FOR THE:
    A. Increased tone of the extensors
    B. Decreased tone of the agonist muscles
    C. Dampening of muscle contraction necessary to avoid overshooting the line
    D. The tremor that is associated with such movements
                                   Ref. p. 685

SECTION IV - PHYSIOLOGY OF THE NERVOUS SYSTEM 159

832. THE ERROR MADE IN THE DEVELOPMENT OF THE FORCE NECESSARY TO COMPLETE A MUSCLE CONTRACTION AIMED AT DOING A CERTAIN WORK IS CALLED:
A. Dystonia
B. Dysynergia
C. Dysmetria
D. Dyslalia
E. Hypotonia
Ref. p. 686

833. ELECTRICAL ACTIVITY OF ONE AREA OF THE CEREBELLUM IS ASSOCIATED WITH ACTIVITY IN A SPECIFIC AREA OF THE CEREBRAL MOTOR CORTEX. THE RELATIONSHIP BETWEEN THESE TWO POINTS (MOTOR CORTEX AND CEREBELLUM) IS SUCH THAT:
A. When the activity in the cerebellum decreases the corresponding area in the motor cortex decreases in activity
B. When the activity in the cerebellum decreases, the corresponding area in the motor cortex is increased in activity
C. When the activity in the cerebellum increases, the corresponding area in the cerebral cortex is decreased in activity
D. There is no predictable correlation between these structures, as other centers are interposed between the cerebellum and motor cortex and modulate each one independently   Ref. p. 687

834. A CEREBELLAR TRAUMA IS MOST LIKELY TO BE ASSOCIATED WITH SEVERE MOTOR DISTURBANCES IF THE LESION INVOLVES 1 CUBIC CENTIMETER OF TISSUE LOCATED IN THE:
A. Cerebellar cortex of the right hemisphere
B. Cerebellar cortex of the left hemisphere
C. Cerebellar cortex of the dorsal area, at the posterior pole
D. Dentate nucleus   Ref. p. 688

835. THE FAILURE OF ONE HAND WHICH TRIES TO TURN ITS PALM UPSIDE AND DOWN IN SYNCHRONY WITH THE OPPOSITE HAND AND WHICH IS NOT ABLE TO KEEP PACE WITH THE NORMALLY MOVING HAND IS CALLED:
A. Dysmetria
B. Ataxia
C. Past pointing
D. Dysdiadochokinesia
E. Athetosis
Ref. p. 688

836. OF ALL THE CEREBELLAR PATHWAYS, ONE OF THE FOLLOWING MAY BE DESTROYED WITHOUT INDUCING INTENTION TREMOR. THESE AFFERENTS REACH THE CEREBELLUM WITH THE:
A. Brachium conjunctivum
B. Dorsal spinocerebellar tracts (from spinal cord)
C. Efferents from the dentate nuclei
D. Efferents from the fastigial nucleus   Ref. p. 688

837. A CEREBELLAR TUMOR IS MOST LIKELY ASSOCIATED WITH SEVERE LOSS OF EQUILIBRIUM WHEN THE LESION INVOLVES:
A. The neocerebellum
B. A small portion of the right hemisphere
C. A small part of the left hemisphere
D. The floculonodular lobe   Ref. p. 688

SECTION IV - PHYSIOLOGY OF THE NERVOUS SYSTEM

838. THE REPETITION OF A DEFINITE SET OF MOVEMENTS LEADING TO A PURPOSEFUL ACT (LIKE COMBING ONE'S HAIR) IS MORE SERIOUSLY HAMPERED BY A SMALL LESION LOCALIZED IN THE:
A. Anterior portion of the motor cortex
B. Premotor cortex
C. Somesthetic cortex
D. Anterior pole of the frontal lobe
E. Superior temporal gyrus   Ref. p. 689

839. THE NAME APPLIED TO POSTULATED PATTERNS OF MOTOR ACTIVITIES APPARENTLY ASSOCIATED THROUGH THE MEDIATION OF THE SOMESTHETIC SENSORY CORTEX IS:
A. Sensory engrams
B. Motor matrices
C. Sensory somatotopia
D. Motor equivalents   Ref. p. 689

840. AS OPPOSED TO SIMPLER MOTOR ACTS, THOSE THAT REQUIRE GREAT SKILL AND VERY RAPID SUCCESSION OF COORDINATED MOVEMENTS ARE BELIEVED TO BE CONTROLLED BY AREAS LOCATED IN THE:
A. Motor areas of the frontal lobe
B. Thalamus
C. Subthalamic nuclei
D. Somesthetic sensory cortex
E. Occipital cortex (areas 18 and 19)   Ref. p. 690

841. AUTONOMIC FIBERS THAT REACH THE SMOOTH MUSCLE OF THE SMALL INTESTINE AND SECRETE NOREPINEPHRINE AT THEIR TERMINAL END PORTION ARE:
A. Originated in neurons which were stimulated by acetylcholine
B. Capable of transmitting afferent stimuli to the cord as well
C. Sometimes cholinergic
D. Originated in neurons located in the intermediolateral horn of the spinal cord   Ref. p. 693

842. IN THE SYMPATHETIC NEURAL CHAIN, THE SO-CALLED GRAY RAMI:
A. Conduct afferent information into the spinal cord
B. Have their end terminal portions secreting acetylcholine in the paravertebral ganglia
C. Pass into spinal nerves and join nerves destined to skeletal muscle and its vessels
D. Are richly myelinated fast fibers
E. Comprise the bulk (about 80%) of the fibers contained in an average skeletal muscle nerve   Ref. p. 693

SECTION IV - PHYSIOLOGY OF THE NERVOUS SYSTEM

843. THE REASON WHY A SYSTEMATIC SURGICAL REMOVAL OF ALL PARAVERTEBRAL GANGLIA CANNOT CAUSE A TOTAL SYMPATHECTOMY IS DUE TO THE FACT THAT:
    A. Some sympathetic fibers emerge with the sacral efferent outflow
    B. Most of the function of the surgically ablated ganglia is compensated for by an increased secretion of the adrenal medulla (epinephrine)
    C. Some preganglionic sympathetic efferent fibers do not synapse in paravertebral ganglia but do so with neurons located within spinal nerves
    D. Surgical procedures should remove dorsal root ganglia in order to obtain a total sympathectomy   Ref. p. 693

844. THE ANATOMICAL DISTRIBUTION OF SYMPATHETIC FIBERS IS SUCH THAT THE VESSELS THAT PERFUSE THE SKIN OF THE FACE HAVE THEIR SYMPATHETIC INNERVATION ORIGINATED IN PREGANGLIONIC NEURONS LOCATED AT THE LEVEL OF SPINAL CORD SEGMENTS:
    A. C_1-C_2
    B. C_3-C_4
    C. T_1-T_2
    D. T_5-T_6
    E. L_1-L_2
    Ref. p. 693

845. THE ADRENAL MEDULLA RECEIVES SYMPATHETIC FIBERS WHOSE NEURONAL CELL BODY IS LOCATED IN:
    A. The celiac ganglion
    B. The paravertebral ganglia next to the 12 thoracic vertebral bodies
    C. A ganglion immediately adjacent to the adrenal gland itself
    D. The spinal cord
    E. Organ of Zuckerkandl   Ref. p. 693

846. THE FIBERS THAT MOVE THE VOCAL CORDS:
    A. Are not contained within the vagus nerve itself
    B. Are part of the voluntary motor system of fibers
    C. Originate in the nodose ganglion
    D. Originate in autonomic ganglia immediately adjacent to the vocal cords where acetylcholine is the neurotransmitter in the synapse present between pre and post-ganglionic neurons
    Ref. p. 693

847. ALL OF THE FOLLOWING ORGANS RECEIVE THEIR PARASYMPATHETIC INNERVATION VIA THE VAGUS NERVES, EXCEPT:
    A. Stomach
    B. Esophagus
    C. Lungs
    D. Cecum
    E. Rectum
    Ref. p. 693

848. IN ALL OF THE FOLLOWING AREAS ACETYLCHOLINE IS THE NEUROTRANSMITTER SUBSTANCE SECRETED, EXCEPT:
    A. Skeletal muscle end plate
    B. Preganglionic endings in adrenal medullary cells
    C. Preganglionic endings in the otic ganglion
    D. Preganglionic endings in the ciliary ganglion
    E. Post-ganglionic sympathetics to Purkinje cells of the heart
    Ref. p. 694

SECTION IV - PHYSIOLOGY OF THE NERVOUS SYSTEM

849. THE MECHANISM BY WHICH ACETYLCHOLINE SECRETED AT POST-GANGLIONIC PARASYMPATHETIC FIBERS IS INACTIVATED IS:
   A. The addition of one extra methyl group to the <u>acetyl</u> (choline) radical
   B. The removal of one methyl group from the choline moeity
   C. The hydrolysis of the ester bond that links the acetyl with the choline group
   D. Through the activity of a methyl transferase
   Ref. p. 694

850. DISAPPEARANCE OF LIBERATED NOREPINEPHRINE IN POST-GANGLIONIC SYMPATHETIC ENDINGS (AFTER ITS RELEASE) OCCURS LARGELY BECAUSE OF:
   A. The action of a transaminase
   B. The transformation into epinephrine
   C. The reabsorption of secreted molecules back into the terminal sympathetic endings
   D. Demethylation
   Ref. p. 694

851. ALL THE FOLLOWING STATEMENTS ARE TRUE, <u>EXCEPT</u>:
   A. Mydriasis is mediated by increased norepinephrine secretion in the post-synaptic endings in the iris
   B. Myosis is mediated by increased acetylcholine secretion in the post-ganglionic parasympathetics in the iris
   C. Accommodation for closer viewing is mediated by increased acetylcholine secreted in the ciliary muscle of the eye
   D. Accommodation for farther distances is mediated by increased norepinephrine secreted in the ciliary muscle of the eye
   Ref. p. 695

852. A MAJOR ROLE OF PARASYMPATHETIC INNERVATION TO VASODILATE ACTIVELY AND TO A MARKED EXTENT OCCURS IN THE AREA OF THE:
   A. Brain
   B. Coronaries
   C. Kidneys
   D. Skin of the dorsum
   E. External genitalia
   Ref. Table 59-1, p. 696

853. INCREASED SWEATING IS MEDIATED BY INCREASED:
   A. Activity of parasympathetic fibers to the skin
   B. Secretion of norepinephrine in end portions of the sympathetic fibers to the skin
   C. Secretion of acetylcholine in the end portions of the parasympathetic fibers to the skin
   D. Secretion of acetylcholine in the post-ganglionic fibers of the sympathetic fibers to the skin
   E. Vasoconstriction due to acetylcholine which in turn releases adenosin-diphosphate and other metabolites
   Ref. p. 696

854. THE INCREASED PERISTALSIS NECESSARY FOR A NORMAL PATTERN OF EMPTYING OF THE LARGE BOWEL REQUIRES A(N):
   A. Increased activity of the sympathetic fibers to the myenteric plexus
   B. Decreased activity of the sympathetic fibers to the myenteric plexus
   C. Increased parasympathetic activity
   D. Decreased parasympathetic activity   Ref. p. 696

SECTION IV - PHYSIOLOGY OF THE NERVOUS SYSTEM   163

855. AN INCREASED SYMPATHETIC ACTIVITY RESULTS IN:
   A. Increased resistance to bronchial air flow
   B. Increased peristalsis
   C. Decreased pulmonary compliance
   D. Decreased pulse velocity
   E. Decreased work of respiration   Ref. p. 697

856. INCREASED ENDOGENOUS SECRETION OF EPINEPHRINE FROM THE ADRENAL MEDULLA IS LIKELY TO CAUSE:
   A. Increased cardiac output and decreased peripheral resistance
   B. Decreased cardiac output and increased peripheral resistance
   C. Decreased cardiac output and decreased peripheral resistance
   D. Decreased cardiac output and decreased work of the heart
                                             Ref. p. 697

857. THE RELEASE OF ENDOGENOUS EPINEPHRINE FROM THE ADRENAL MEDULLA THAT FOLLOWS A STRESS RESULTS IN EFFECTS WHICH ARE SLIGHTLY DIFFERENT FROM THOSE THAT CAN BE OBTAINED IN THE ADRENALECTOMIZED ANIMAL BY STIMULATING THE SYMPATHETIC SYSTEM THROUGH A SIMILAR STRESS. THUS, WHEN COMPARING RELATIVE EFFECTS OF EPINEPHRINE VERSUS NOREPINEPHRINE, IT SHOULD BE REMEMBERED THAT EPINEPHRINE, RELEASED ENDOGENOUSLY, TENDS TO CAUSE:
   A. A lesser effect on the work of the heart
   B. A more severe diastolic and systolic hypertension
   C. A more severe arteriolar vasoconstriction
   D. Effects that last longer
   E. A lesser danger of causing ventricular fibrillation
                                             Ref. p. 698

858. AN ANIMAL DEPRIVED OF ITS ADRENAL GLANDS:
   A. Loses most of the reactions due to increased sympathetic activity, because of the lack of circulating epinephrine
   B. Can still develop tachycardia and sympathetic hypertension in response to exercise
   C. Has a much greater rate of glycogenolysis
   D. Usually is hyperglycemic during fasting periods
                                             Ref. p. 698

859. A PHYSIOLOGICAL ACTION OF EPINEPHRINE BELIEVED TO BE DUE TO CONTACT WITH SO-CALLED "BETA RECEPTORS" IS THAT WHICH IS MANIFESTED BY:
   A. Vasoconstriction in skeletal muscle
   B. Mydriasis
   C. Intestinal relaxation
   D. Contraction of pilomotor smooth muscle
   E. Increased ventricular contractility   Ref. Table 59-2, p. 698

860. A BASAL STATE OF TONIC FIRING OF PREGANGLIONIC SYMPATHETIC FIBERS OCCURS AT A RATE EQUAL TO:
   A. 1-2 per second
   B. 20-30 per second
   C. 50-100 per second
   D. 500-1000 per second   Ref. p. 698

SECTION IV - PHYSIOLOGY OF THE NERVOUS SYSTEM

861. A SUDDEN DECREASE IN THE RATE OF FIRING FROM THE BASAL LEVEL, SEEN IN TONIC ACTIVITY OF THE SYMPATHETIC SYSTEM INNERVATING A VASCULAR PLEXUS, RESULTS IN:
    A. No change in vascular diameter
    B. A myogenic-induced constriction
    C. A decreased volume of blood flowing through the capillary system
    D. A significant vasodilation in such territory
    E. An accumulation of tissue metabolites
    Ref. p. 699

862. VAGOTOMY RESULTS IN:
    A. More rapid emptying of the gastric chyme in the duodenum
    B. Hyperperistaltic activity of the bowel
    C. No change in gastrointestinal motility
    D. A quiescent stomach
    Ref. p. 699

863. FROM ABLATION EXPERIMENTS WE MAY CONCLUDE THAT:
    A. Vagotomy may cause the heart rate to decrease significantly
    B. Most of the arteriolar vasoconstriction present during sleep is due to a tonic rate of firing of post-ganglionic sympathetic fibers and not to adrenal secretion
    C. Most of the arteriolar vasoconstriction present at rest is due to a basal secretion of norepinephrine and epinephrine
    D. No epinephrine is secreted by adrenal medulla except during emergencies
    Ref. p. 699

864. IN AN ESOPHAGEAL DISEASE WHERE THE MYENTERIC PLEXUS IS DESTROYED, THE PARENTERAL INJECTION OF ACETYLCHOLINE (OR A DERIVATIVE) SHOULD BE EXPECTED TO:
    A. Accelerate the heart
    B. Have no effect
    C. Diminish an overactive hyperperistaltism
    D. Cause a violent spasm of the smooth muscle
    E. Dilate the bronchioles
    Ref. p. 699

865. NOREPINEPHRINE ADMINISTERED TO A PERSON WHOSE SYMPATHETIC SYSTEM HAS BEEN SURGICALLY REMOVED REACTS BY DEVELOPING:
    A. A bradycardia
    B. A more marked sympathetic response than before sympathectomy
    C. Hypoglycemia
    D. Hypotension
    Ref. p. 699

866. THE INCREASED SECRETION OF GASTRIC GLANDS APPEARING JUST BEFORE EATING AN APPETIZING FOOD WHOSE SIGHT OR SMELL HAS BEEN PERCEIVED:
    A. Can be inhibited by blocking beta adrenergic receptors
    B. Can be inhibited by blocking alpha adrenergic receptors
    C. May be totally suppressed by a bilateral vagotomy
    D. Is a purely hormonal effect
    Ref. p. 700

SECTION IV - PHYSIOLOGY OF THE NERVOUS SYSTEM 165

867. IN ALL OF THE FOLLOWING ACTS, A REFLEX ARC IS INVOLVED, WHERE THE AFFERENT OR THE EFFERENT ARE PART OF THE AUTONOMIC NERVOUS SYSTEM, EXCEPT:
   A. Consensual pupillary constriction as a reaction to light of increased intensity
   B. Emptying of the rectum
   C. Emptying of the lactating female mammary gland in response to sucking by the infant
   D. Emptying of the urinary bladder
   E. Sweating Ref. pp. 700, 988-989

868. IT IS BELIEVED THAT THE MECHANISM THROUGH WHICH EPHEDRINE (A SYMPATHOMIMETIC AGENT) ACTS, IS ITS CAPACITY TO:
   A. Combine with beta receptors
   B. Combine with alpha receptors
   C. Combine with gamma receptors
   D. Release norepinephrine from sympathetic fibers
   E. Inhibit the metabolism of norepinephrine once it is released by sympathetic fibers Ref. p. 701

869. THE PHARMACOLOGICAL AGENT BELIEVED TO ACT BY INHIBITION OF SYNTHESIS AND STORAGE OF NOREPINEPHRINE IS:
   A. Reserpine
   B. Guanethedine
   C. Dibenamine
   D. Hexamethonium
   E. Pentobarbital
   Ref. p. 701

870. PARASYMPATHOMIMETIC DRUGS ADMINISTERED IN PHARMACOLOGICAL DOSES:
   A. Do not cause sweating
   B. Act selectively on cranial and sacral parasympathetic branches but not in areas innervated by the sympathetic cholinergic fibers
   C. Are synergistic in their action with drugs which inhibit cholinesterase (such as neostigmine)
   D. Do not have an effect on the eye if administered by injection
   Ref. p. 701

871. ATROPINE BLOCKS THE EFFECTS OF ACETYLCHOLINE IN RECEPTORS LOCATED IN ALL THE AREAS LISTED BELOW, EXCEPT:
   A. The iris
   B. The Auerbach plexus
   C. The atrioventricular node
   D. The skeletal muscle end plate
   E. Mucous glands in the bronchial mucosa
   Ref. p. 701

872. NICOTINE HAS AN EXCITATORY INFLUENCE ON:
   A. Receptors normally responsive to postganglionic parasympathetic fibers
   B. Receptors located in neurons present in sympathetic ganglia
   C. Beta-adrenergic receptors
   D. Alpha-adrenergic receptors
   E. The sympathetic paravertebral ganglia but not on parasympathetic ganglia Ref. p. 702

SECTION IV - PHYSIOLOGY OF THE NERVOUS SYSTEM

873. THE SUCCESSFUL USE OF SO-CALLED GANGLIOPLEGIC DRUGS (E.G. HEXAMETHONIUM) IN TREATING PATIENTS AFFLICTED BY ARTERIAL HYPERTENSION IS DUE TO THE:
    A. Preferential inhibition of synaptic transmission on sympathetic ganglia
    B. Exclusive inhibition of sympathetic ganglia transmission, without any effect on parasympathetic ganglia
    C. Effect of such drugs on the alpha receptors in smooth muscle of arterioles
    D. Effect of such drugs on the beta receptors in smooth muscles of arterioles
    Ref. p. 702

874. AN INCREASE IN THE TEMPERATURE OF THE BLOOD THAT PERFUSES THE HYPOTHALAMUS WILL SPECIFICALLY ACTIVATE TO A GREATER DEGREE NEURONS LOCATED IN THE:
    A. Mamillary body
    B. Dorsomedial nucleus
    C. Median eminence
    D. Ventromedial nucleus
    E. Anterior hypothalamus
    Ref. Fig. 59-6, p. 703

875. SELECTIVE DAMAGE TO THE LATERAL HYPOTHALAMIC AREA IS LIKELY TO RESULT IN:
    A. Overeating
    B. Diabetes insipidus
    C. Anorexia
    D. Thirst
    E. Lack of libido
    Ref. p. 703

876. EXTREMELY OBESE HUMANS WHOSE DISEASE CAN BE ASCRIBED TO A HYPOTHALAMIC LESION ARE LIKELY TO HAVE DECREASED NEURONAL FUNCTION IN THE AREA OF THE:
    A. Ventromedial nucleus
    B. Mamillary bodies
    C. Preoptic area
    D. Lateral area
    E. Supraoptic nuclei
    Ref. p. 704

877. A PERSON WHO HAS SUFFERED A TRAUMATIC DAMAGE TO A PORTION OF THE FACILITATORY AREA OF THE RETICULAR FORMATION IN THE BRAIN STEM IS LIKELY TO SHOW AMONG OTHER THINGS:
    A. Coma
    B. Euphoria
    C. Alexia
    D. Blindness
    E. Anosmia
    Ref. p. 705

878. THE COMPONENT OF ASCENDING FIBERS WHICH DEPARTS FROM THE RETICULAR FORMATION IN THE BRAIN STEM AND PASSES THROUGH THE MIDLINE NUCLEI OF THE THALAMUS BEFORE REACHING THE CORTEX, IS ASSOCIATED WITH:
    A. Immediate arousal
    B. Long lasting inhibition of the motor cortex
    C. Sleep
    D. Activation of all parts of the cerebral cortex to the same extent
    E. Activation of selective areas of the cortex in preference to other portions which are stimulated to a lesser extent
    Ref. pp. 705-706

SECTION IV - PHYSIOLOGY OF THE NERVOUS SYSTEM 167

879. AN ANIMAL MAY RESPOND BY DEVELOPING TYPICAL ELECTRO-
ENCEPHALOGRAPHIC EVIDENCE OF SLEEP IF:
A. Its reticular formation is stimulated in an area located in the upper medulla and lower pons
B. It is exposed to loud noises
C. It is forced to move around
D. Its reticular formation is stimulated in the area of the cerebral peduncles and subthalamic nuclei   Ref. p. 708

880. RAPID EYE MOVEMENTS (REM) ARE PRESENT DURING TRANSIENT PHASES CALLED REM SLEEP. DURING REM SLEEP IT IS TYPICAL TO FIND THAT THE ANIMAL HAS:
A. A lower blood pressure and bradycardia
B. A high cardiac output
C. Increased muscle tone
D. An electroencephalogram which suggests a very deep state of sleep
E. A high diastolic blood pressure   Ref. p. 709

881. IN A PERSON WHO SLEEPS WELL COVERED, IN A ROOM AT NORMAL TEMPERATURE (75-77° F), SEVERAL PHYSIOLOGICAL CHANGES TAKE PLACE WHICH ARE NOT USUALLY FOUND DURING THE WAKING STATE, BUT WHICH ARE CHARACTERISTIC OF THE NORMAL SLEEP-ING STATE. THUS, ONE WOULD NOT EXPECT TO SEE DURING NORMAL SLEEP A:
A. Lower arterial blood pressure than awake
B. Decreased heart rate
C. Cutaneous vasoconstriction
D. Diminished tone in the skeletal muscles
E. Diminished metabolic rate   Ref. p. 709

882. THERE IS EVIDENCE THAT MAJOR PATHWAYS CONNECTING ONE CEREBRAL HEMISPHERE WITH THE OPPOSITE SIDE, AND ONE AREA OF THE BRAIN (MOTOR) WITH ANOTHER (SENSORY) EXIST AT THE LEVEL OF THE:
A. Corpus callosum
B. Septum pellucidum
C. Anterior commissure
D. Posterior commissure
E. Reticular formation
Ref. p. 710

883. A PERSON WHO IS AWAKE AND MENTALLY RELAXED, AND NOT CON-CENTRATED ON ANY PARTICULAR SUBJECT OR OBJECT, IS LIKELY TO SHOW IN THE ELECTROENCEPHALOGRAM A PREDOMINANCE OF:
A. Alpha waves
B. Beta waves
C. Theta waves
D. Delta waves   Ref. p. 711

884. A SUDDEN TEMPORARY MENTAL FRUSTRATION WILL GIVE RISE, IN THE ELECTROENCEPHALOGRAPHIC TRACING, TO A BURST OF:
A. Alpha waves
B. Beta waves
C. Theta waves
D. Delta waves   Ref. p. 712

SECTION IV - PHYSIOLOGY OF THE NERVOUS SYSTEM

885. A NEUROSURGICAL ABLATION OF A PORTION OF THE BRAIN TISSUE SUBJACENT TO AN AREA OF THE CEREBRAL CORTEX WILL CAUSE, IN THE AREA ABOVE IT, THE APPEARANCE OF WAVES WITH A FREQUENCY EQUAL TO:
   A. 10 per second
   B. 50 per second
   C. 100 cycles per second
   D. 1 cycle every 2 or 3 seconds Ref. p. 712

886. DELTA WAVES CAN OCCUR IN THE ELECTROENCEPHALOGRAPHIC RECORD OF A:
   A. Normal person who is awake and who has no neurological deficit
   B. Normal person who is asleep in the REM type of sleep
   C. Person who sees a bright light
   D. Soundly sleep (non REM) child Ref. p. 712

887. STIMULATION OF THE THALAMIC PORTION OF THE RETICULAR FORMATION RESULTS IN:
   A. Monosynaptic brief immediate responses in the whole cortex
   B. Theta waves
   C. Beta waves
   D. Slowly developed, long lasting, electrical potentials in specific localized areas of the cerebral cortex Ref. p. 712

888. IT IS POSSIBLE TO INDUCE THE APPEARANCE OF AN ABNORMAL ELECTROENCEPHALOGRAM BY:
   A. Having the patient close his eyes
   B. Asking the patient to perform arithmetical calculations
   C. Tilting the patient to the supine position
   D. Inducing alkalosis by forced voluntary hyperventilation Ref. p. 713

889. THE ORIGIN OF THE ABNORMALLY EXCESSIVE NEURONAL DISCHARGE SEEN IN GRAND MAL EPILEPSY IS THE:
   A. Spinal cord
   B. Brain stem medial lemniscus
   C. Brain stem reticular formation
   D. Hypothalamus
   E. White matter of the cerebral cortex in the frontal lobe Ref. p. 714

890. IT IS TYPICAL OF THE ELECTROENCEPHALOGRAPHIC TRACING INSCRIBED DURING A SEIZURE OF GRAND MAL EPILEPSY THAT:
   A. The occipital lobes are silent
   B. The motor cortex has focal areas of high activity while the temporal lobes are almost inactive
   C. The frequency of discharge is relatively slow and synchronous
   D. High voltage discharge occurs throughout the cortex
   E. It waxes and wanes with respirations Ref. p. 714

SECTION IV - PHYSIOLOGY OF THE NERVOUS SYSTEM

891. ATYPICAL OF WHAT IS KNOWN AS PETIT MAL EPILEPSY (WOULD NOT USUALLY BE ASSOCIATED WITH THIS FORM OF EPILEPSY AND IF MANIFESTED WOULD SERVE TO RULE OUT THE DIAGNOSIS OF PETIT MAL) IS:
    A. An occasional myoclonus
    B. Brief periods (10-20 seconds) of unconsciousness
    C. Blinking of the eyes during the attack
    D. A spike and dome electroencephalographic pattern
    E. Clonic convulsions lasting 5 minutes   Ref. p. 714

892. A "JACKSONIAN" EPILEPTIC SEIZURE IS CHARACTERIZED BY:
    A. A spike and dome electroencephalographic pattern
    B. An abnormally violent attack of rage
    C. A very fast, very high voltage electroencephalographic pattern
    D. Muscular contractions that involve one part of an extremity and then subsequently other parts of the same extremity in a progressive fashion
    E. Total absence of motor disturbances but severe amnesia
    Ref. p. 715

893. THE PRIMARY SENSORY CORTEX IS PREDOMINANTLY COMPOSED OF:
    A. Granular cells
    B. Pyramidal cells
    C. Fusiform cells
    D. Polymorphic cells
    E. Betz cells
    Ref. p. 717

894. CORTICO-THALAMIC CONNECTIONS ARE ORGANIZED SO THAT EACH AREA OF THE CORTEX SENDS AND RECEIVES INFORMATION TO AND FROM A DEFINITE THALAMIC NUCLEUS. THE <u>MOTOR</u> CORTEX IS RELATED TO THE THALAMIC NUCLEUS KNOWN AS:
    A. Postero-lateral nucleus
    B. Lateral geniculate body
    C. Dorso-medial nucleus
    D. Ventrolateral nucleus
    E. Medial geniculate body
    Ref. p. 717

895. AFTER A PRIMARY SENSATION (SUCH AS VISION) HAS STIMULATED THE "PRIMARY" VISUAL CORTEX THE PROCESS OF <u>INTERPRETATION</u> FOLLOWS. THE LACK OF INTERPRETATION OF THE MEANING OF A GIVEN LETTER, WHICH CAN BE SEEN BUT NOT UNDERSTOOD, IS MOST LIKELY TO RESULT IF A TUMOR OR OTHER DESTRUCTIVE LESION DAMAGES AN AREA OF CEREBRAL CORTEX LOCATED IN THE:
    A. Temporal lobe, rostral portion of the superior and middle gyri
    B. Parietal lobe, immediately caudal to the somesthetic area
    C. Areas 18 and 19 of the occipital lobe
    D. Frontal lobe, rostral to the motor area on the dominant side
    E. Cyngulate gyrus   Ref. p. 719

896. THE GREATEST COMPLEXITY OF THOUGHTS AND MEMORY PATTERNS CAN BE OBTAINED BY ELECTRICALLY STIMULATING ON THE DOMINANT SIDE:
    A. Areas 18 and 19
    B. The somesthetic cortex (somatic area I)
    C. The corpus callosum
    D. The angular gyrus
    E. The frontal lobe (anterior pole)   Ref. p. 719

SECTION IV - PHYSIOLOGY OF THE NERVOUS SYSTEM

897. THE AREA DESCRIBED IN THE PRECEDING QUESTION IS LOCATED:
   A. Immediately rostral to area 17 (visual cortex)
   B. At the foot of the ascending parietal gyrus
   C. In the rostral portion of the corpus callosum
   D. Where the temporal, parietal and occipital lobes join together
   E. The most rostral portion of the dominant hemisphere
   Ref. p. 719

898. SEVERE MEMORY LOSS FOR RECENTLY ACQUIRED IMPRESSIONS WOULD FOLLOW THE PATHOLOGICAL DESTRUCTION OF THE CEREBRAL CORTEX OF THE MAJOR PORTION OF THE:
   A. Frontal labe
   B. Insula (of Reil)
   C. Occipital cortex
   D. Parietal cortex
   E. Temporal cortex
   Ref. p. 720

899. ASSUME A MONKEY IS TAUGHT TO PERFORM IN A SPACE CAPSULE A SERIES OF FAIRLY COMPLICATED MOVEMENTS. THIS ACQUIRED ABILITY WILL MOST LIKELY BE LOST IF A LESION OF THE BRAIN INVOLVES THE:
   A. Septum pelucidum
   B. Ventromedial nucleus of the hypothalamus
   C. Areas of the frontal lobes immediately rostral to the motor areas
   D. Area of the parietal lobe immediately caudal to the somesthetic cortex
   E. Auditory cortex
   Ref. p. 720

900. THE CAPACITY OF AN ANIMAL TO CONCENTRATE ON A TASK DEPENDS TO A VERY GREAT EXTENT ON AN INTACT:
   A. Frontal lobe
   B. Parietal lobe
   C. Temporal lobe
   D. Occipital lobe
   E. Hypothalamus
   Ref. p. 721

901. A HEAD INJURY SUSTAINED BY HITTING THE HEAD AGAINST THE FRAME OF A CAR IN AN AUTOMOBILE COLLISION IS LIKELY TO:
   A. Decrease recent memory
   B. Cause euphoria
   C. Suppress color vision
   D. Improve the capacity to retain memory engrams immediately after the collision
   Ref. p. 722

902. IT IS BELIEVED THAT PART OF THE FUNCTIONAL ROLE OF THE CORPUS CALLOSUM IS RELATED TO:
   A. Conveying part of the motor activity from one hemisphere to another
   B. Serving as another of several pathways for bilateral acoustic representation
   C. Bilateral extraocular muscular synergism
   D. Impress memory engrams bilaterally
   E. Distribute thalamic radiations to the cerebral cortex
   Ref. p. 724

903. A PATIENT, WHO, AFTER A "STROKE," IS ABLE TO WRITE HIS THOUGHTS TO HIS FAMILY BUT IS UNABLE TO EXPRESS THESE IDEAS AS SPOKEN WORDS IS AFFECTED BY:
   A. Motor aphasia
   B. Sensory aphasia
   C. Visual agnosia
   D. A case of word deafness
   E. Apraxia
   Ref. p. 727

SECTION IV - PHYSIOLOGY OF THE NERVOUS SYSTEM 171

904. ANOTHER PATIENT WHO UNDERSTANDS THE MEANING OF IDEAS SPOKEN TO HIM BUT IS UNABLE TO FORMULATE HIS IDEAS COHERENTLY IS AFFECTED BY:
A. A motor aphasia
B. A sensory aphasia
C. A visual agnosia
D. A case of word deafness
E. Apraxia
Ref. p. 727

905. THE VENTROMEDIAL NUCLEI OF THE HYPOTHALAMUS ARE THE SITE OF LOCATION OF NEURONS WHICH SEEM TO ORIGINATE:
A. Defensive sensations
B. Agressive behavior
C. Feeding movements
D. Sensations of "reward"
E. Headache
Ref. p. 730

906. THE PERIFORNICAL NUCLEUS OF THE HYPOTHALAMUS SEEMS TO CONVEY:
A. Pleasure
B. "Reward"
C. Feeding desires
D. Sexual drive
E. Pain and unpleasant feelings
Ref. p. 730

907. IT SEEMS THAT EXPERIENCES WHICH ARE LEAST REMEMBERED ARE THOSE THAT EVOKE:
A. Pleasure or pain
B. Neither pleasure or punishment
C. Attention
D. Reward
Ref. p. 731

908. THE REACTION OF RAGE EVOKED IN AN EXPERIMENTAL ANIMAL:
A. Requires that its frontal lobe be intact
B. Requires that the sensory cortex be connected to the thalamus
C. Requires that memory of previous events be complete
D. Can be obtained in a decorticated animal
Ref. p. 731

909. OF MANY AREAS OF THE HYPOTHALAMUS, ONE LIKELY TO CAUSE SLEEP IF STIMULATED ELECTRICALLY IN AN EXPERIMENTAL ANIMAL, IS THE AREA LOCATED IN THE:
A. Mamillary bodies
B. Perifornical nuclei
C. Anterior hypothalamus
D. Median eminence
Ref. p. 731

910. THE HYPOTHALAMIC AREAS MOST DIRECTLY RELATED TO EXCITING THE SYMPATHETIC NERVOUS SYSTEM ARE THE:
A. Mamillary bodies
B. Perifornical nuclei
C. Anterior hypothalamus
D. Median eminence
E. Dorsal areas
Ref. p. 731

911. A COMPLEX OF SEXUALLY ORIENTED MOVEMENTS, INCLUDING OVULATION, CAN BE OBTAINED BY ELECTRICAL STIMULATION OF CERTAIN AREAS OF THE:
A. Flocculo-nodular areas of the cerebellum
B. Anterior pole of the frontal lobes (rostral to premotor areas)
C. Posterior pole of the occipital lobes (visual cortices)
D. Anterior pole of the temporal lobes (amygdala)
E. Dorsal surface of the parietal lobes (somesthetic association areas)
Ref. p. 733

SECTION IV - PHYSIOLOGY OF THE NERVOUS SYSTEM

912. A PERSON UNDERGOING NEUROSURGERY MAY, WHILE CONSCIOUS, BE STIMULATED WITH WEAK ELECTRICAL IMPULSES ON THE CORTEX OF VARIOUS PARTS OF THE BRAIN. VISUAL, OLFACTORY AND OTHER HALLUCINATIONS CAN BE EASILY OBTAINED IF THE STIMULUS IS APPLIED TO THE CORTEX OF THE:
A. Frontal lobe, motor cortex
B. Frontal lobe, premotor area
C. Parietal lobe, somesthetic cortex
D. Occipital lobe, visual cortex
E. Hyppocampus

Ref. p. 733

SECTION V - GASTROINTESTINAL PHYSIOLOGY

FOR EACH OF THE FOLLOWING MULTIPLE CHOICE QUESTIONS, SELECT THE ONE MOST APPROPRIATE ANSWER:

913. WHEN THE SYMPATHETIC SYSTEM DEVELOPS AN INCREASED LEVEL OF ACTIVITY THERE IS AS A RESULT A(N):
    A. Increased peristaltism in the sigmoid colon
    B. Decreased tone of the ileocecal sphincter
    C. Increased tone of the external anal sphincter
    D. Increased tone of the internal anal sphincter
    E. Reversed peristaltism of the sigmoid colon
    Ref. p. 740

914. DEFECATION MAY BE SEVERELY IMPAIRED IF THE:
    A. Pelvic sympathetic plexus is excised surgically
    B. Sympathetic plexus in the pelvis is stimulated by pharmacological means
    C. Pelvic parasympathetic nerves are stimulated
    D. Myenteric plexus (Auerbach) is destroyed by disease
    Ref. p. 740

915. THE FOLLOWING EVENTS OCCUR IN A NORMAL SWALLOWING SEQUENCE, EXCEPT THE:
    A. Palatopharyngeal folds approximate to each other
    B. Epiglottis covers the upper portion of the larynx which is moved upward
    C. Vocal cords relax and separate from each other
    D. Hypopharyngeal muscle becomes relaxed
    E. Superior pharyngeal constrictor contracts
    Ref. pp. 742-743

916. A LARGE PROPORTION OF AFFERENT SENSORY INFORMATION, WHICH INITIATES THE ORDERLY SEQUENCE OF MUSCLE CONTRACTION TYPICAL OF AN ACT OF SWALLOWING, ASCENDS TO THE CENTRAL NERVOUS SYSTEM WITH THE:
    A. Vagus
    B. Glossopharyngeus
    C. Facial
    D. Hypoglossus
    E. Intermedius
    Ref. p. 743

917. OF THE FOLLOWING CRANIAL NERVES, ONE DOES NOT PARTICIPATE WITH MOTOR EFFERENTS TO THE ACT OF SWALLOWING:
    A. Trigeminal nerve
    B. Glossopharyngeus
    C. Accessory (XI cranial nerve)
    D. Hypoglossus
    E. Intermedius
    Ref. p. 743

918. DURING A NORMAL SWALLOWING SEQUENCE, THE SMOOTH MUSCLE WHICH SURROUNDS THE GASTROESOPHAGEAL JUNCTION:
    A. Remains tonically constricted until the bolus arrives at this junction
    B. Relaxes even before the oncoming peristaltic wave has arrived
    C. Constricts further on the passing bolus and allows a gradual esophageal emptying
    D. Is never constricted at all and has no participation of any kind in the swallowing process
    Ref. p. 744

SECTION V - GASTROINTESTINAL PHYSIOLOGY

919. THE REPETITIVE PERISTALTIC WAVES SWEEPING THE LENGTH OF THE STOMACH AFTER A MEAL:
   A. May increase the intracavitary pressure by as much as 50 mm Hg
   B. Never increase the intragastric pressure due to the "plasticity" of the stomach
   C. Are more powerful in the fundus and tend to decrease significantly in force as they reach the antrum
   D. Are not related to the emptying of the stomach but are only concerned with mixing gastric contents   Ref. p. 745

920. SO-CALLED HUNGER "PANGS":
   A. Are truly due to smooth muscle spasm of the transverse colon
   B. Are associated with esophageal and pyloric contractions that coexist with an empty relaxed stomach
   C. Are more common if starvation is associated with hypoglycemia
   D. May occur in a totally vagotomized patient
       Ref. p. 746

921. GASTRIC EMPTYING IS NOT DIRECTLY INFLUENCED BY:
   A. The intragastric pressure developed by the peristaltic waves reaching the pylorus
   B. The acidity of the chyme reaching the first portion of the duodenum
   C. The osmolarity of the gastric chyme entering the duodenum
   D. Enterogastrone
   E. Secretin   Ref. p. 747

922. A MAJOR PART OF GALLBLADDER CONTRACTIONS ARE DUE TO:
   A. Sympathetic contraction of the viscus
   B. The overfilling of the gallbladder with bile
   C. The rate of cholesterol synthesis and excretion by the liver
   D. A hormone synthesized by the duodenal mucosa
   E. A pancreatic hormone   Ref. p. 749

923. THE PASSAGE OF ILEAL CONTENTS INTO THE CECUM THROUGH THE ILEOCECAL VALVE IS ACCELERATED BY:
   A. Distension of the cecum with increased pressure within the cecum
   B. Inflammation of the area of the cecum
   C. Peritonitis of the region around the cecum and appendix
   D. Renal colic
   E. Ingestion of food   Ref. p. 749

924. IN THE SEQUENCE OF EVENTS KNOWN AS DEFECATION:
   A. The myenteric plexus of the rectal portion of the large intestine gives rise to new peristaltic waves in the sigmoid even without the participation of the sacral segment of the spinal cord
   B. A normally functional spinal cord is essential for a successful emptying of the rectum
   C. The sigmoid colon could not contract if the nervi erigentes are cut
   D. The external sphincter must relax by stimulation of the parasympathetic pelvic nerves   Ref. p. 751

SECTION V - GASTROINTESTINAL PHYSIOLOGY 175

925. THE ARRIVAL OF PARASYMPATHETIC NERVOUS IMPULSES TO THE BASE OF SECRETORY CELLS IN THE UPPER PORTION OF THE GASTROINTESTINAL TRACT (MOUTH, STOMACH):
    A. Decreases the normal membrane potential (makes the interior of the cell less negative)
    B. Makes the interior of the cell positive with respect to the extracellular fluid
    C. Causes changes in electrical potential that correlate with the possibility that anions increase in concentration inside the cell
    D. Causes changes that are much less significant than those induced by stimulation of sympathetic nerves   Ref. p. 755

926. MUCUS PRESENT ON THE SURFACE OF THE MUCOSA OF THE STOMACH, SMALL AND LARGE INTESTINE IS:
    A. Resistant to enzymes secreted into the lumen of the gastrointestinal tract
    B. Alkaline and thus a poor buffer for alkaline foods
    C. A non-amphoteric protein
    D. Easily removed from the surface of the stomach by the gastric secretions during the gastric phase of digestion
       Ref. p. 755

927. THE ALPHA-AMYLASE PRESENT IN THE SECRETORY PRODUCTS OF THE PAROTID GLAND HAS AN OPTIMAL ACTIVITY:
    A. That allows it to act in the stomach two hours after the food has passed into the stomach
    B. Around 6.9
    C. Around 8.0
    D. Very different from that found in the pancreatic amylase
    E. Around 4.5   Ref. p. 756

928. SALIVATION CAN BECOME A CONDITIONED REFLEX. THIS SUGGESTS THAT:
    A. Pleasant taste sensations are not related to the reflex
    B. Only salivatory nuclei in the brain stem need to be excited by taste sensations without participation of suprasegmental influences
    C. The cerebral cortex partially controls salivation
    D. Salivation could be completely interrupted in a decorticate animal whose tongue is mechanically stimulated
       Ref. p. 756

929. IT IS BELIEVED THAT IN PARIETAL CELLS OF THE GASTRIC MUCOSA-WHICH SECRETE HYDROCHLORIC ACID-AN ACTIVE TRANSPORT SYSTEM IS RESPONSIBLE FOR THE TRANSMEMBRANE PASSAGE OF:
    A. Chloride
    B. Carbon dioxide
    C. Bicarbonate
    D. Water   Ref. p. 757

SECTION V - GASTROINTESTINAL PHYSIOLOGY

930. INHIBITION OF CARBONIC ANHYDRASE IN THE GASTRIC MUCOSA LEADS TO VIRTUAL SUPPRESSION OF THE SECRETION OF HYDROCHLORIC ACID. THIS HAS LED TO THE ASSUMPTION THAT:
    A. Hydrochloric acid results from the dissociation of a carboxylic acid
    B. Work required for the secretion of hydrochloric acid is insignificant
    C. Hydration of carbon dioxide plays a critical role in the formation of hydrochloric acid
    D. Achlorhydria is the result of endogenous formation of carbonic anhydrase inhibitors   Ref. p. 757

931. PEPSIN, SECRETED BY THE GASTRIC MUCOSA:
    A. Continues to act in the neutral medium of the duodenum
    B. Is proteolytic on the mucosal surface, at the alkaline medium provided by the mucus secreted by the surface cells of the stomach
    C. Is synthesized within the chief cells
    D. Does not increase in concentration in response to the mental stimuli found in the so-called cephalic phase of gastric digestion
        Ref. pp. 757-759

932. IT IS KNOWN THAT GASTRIN:
    A. Is a large protein molecule, somewhat similar in size to pepsin
    B. Is not secreted by the empty stomach when peristaltic movements may be quite forceful
    C. Reaches the secretory cells of the fundus of the stomach through the blood and not through the lumen
    D. Promotes the secretion of pepsin but not that of hydrochloric acid
        Ref. p. 758

933. THE SECRETION OF GASTRIN BY THE PYLORIC ANTRUM CEASES:
    A. When the stomach is distended by a full meal
    B. When the gastric contents have a pH of about 2.0
    C. If histamine is injected
    D. If the vagi are stimulated   Ref. p. 759

934. UNDER NORMAL CONDITIONS, THE BULK OF GASTRIC SECRETIONS ARE PRODUCED:
    A. Before the food is ingested and while the pleasure of its ingestion is anticipated
    B. While the food stays in the stomach
    C. After the food has entered the duodenum
    D. During fasting periods   Ref. p. 759

935. TRYPSINOGEN, THE INACTIVE PROTEOLYTIC ENZYME SYNTHESIZED BY PANCREATIC ACINAR CELLS:
    A. Is normally activated by enterokinase in the intestinal lumen
    B. Is transformed into chymotrypsin, a shorter and active molecule
    C. Competes with carboxypeptidase for its substrate
    D. Is frequently activated by bile within pancreatic canaliculi
        Ref. p. 760

SECTION V - GASTROINTESTINAL PHYSIOLOGY

936. AN ACUTE PANCREATITIS:
 A. Is frequently a mild disease that causes flatulence
 B. Will lead to insufficient fat digestion and malabsorption if the patient survives
 C. Is genetically induced by the insufficient synthesis of a trypsin inhibitor
 D. Can usually be arrested if proper therapy is applied to stop the necrotizing process   Ref. p. 760

937. IT IS KNOWN THAT SECRETIN:
 A. Is a large protein hormone synthesized by the pancreas, together with pancreozymin
 B. Is a small polypeptide synthesized by the intestinal mucosa
 C. Neutralizes directly the acid chyme that passes through the pylorus
 D. Has an optimal activity at a pH equal to 8.4
           Ref. p. 761

938. THE MAJOR ANION PRESENT IN PANCREATIC SECRETIONS OBTAINED THROUGH THE STIMULATORY EFFECT OF SECRETIN IS:
 A. Chloride   D. Bicarbonate
 B. Bromide   E. Phosphate
 C. Protein   Ref. p. 761

939. THE MAJOR FACTOR THAT STIMULATES THE RELEASE OF SECRETIN INTO THE BLOOD STREAM IS:
 A. An acid pH of the chyme entering the duodenum
 B. Parasympathetic stimuli
 C. Peptones in the gastric chyme that enters the duodenum
 D. A stomach full of digested contents   Ref. p. 761

940. THE PREDOMINANCE OF PANCREATIC ENZYMES OVER THAT OF WATER AND IONS IN INCREASED QUANTITIES OF PANCREATIC SECRETIONS IS ALSO KNOWN BY THE NAME:
 A. Hydrelatic
 B. Ecbolic
 C. Pancreozymic
 D. Diastasic   Ref. p. 761

941. THE TYPE OF PANCREATIC SECRETION REFERRED TO IN THE PREVIOUS QUESTION IS BELIEVED TO BE INDUCED BY:
 A. A hormone synthesized in the duodenal mucosa
 B. A chyme with a low pH
 C. Secretin
 D. A very large gastric distension   Ref. p. 761

942. DURING THE PHASE THAT ANTICIPATES THE ACTUAL INGESTION OF FOOD (SMELL, APPETITE, PLEASANT ANTICIPATION), THE PANCREAS:
 A. Does not eject any secretion into the duodenal lumen
 B. Secretes large amounts of enzymes into the duodenum
 C. Releases small quantities of secretion with enzymes into the duodenal lumen
 D. Receives no signals through the autonomic nerves
           Ref. p. 761

SECTION V - GASTROINTESTINAL PHYSIOLOGY

943. THE FOLLOWING FACTORS PROMOTE BILIARY SECRETION BY THE LIVER, EXCEPT:
   A. Stimulation of parasympathetic fibers reaching the liver
   B. Increase in concentration of bile salts in blood during the enterohepatic circulation of bile salts
   C. Secretin formed in the duodenal mucosa
   D. Pancreozymin secretion
   E. Increase in hepatic blood flow   Ref. p. 762

944. MUCUS SECRETION BY THE DUODENAL BRUNNER'S GLANDS ARE BELIEVED TO BE INHIBITED BY:
   A. Stimulation of the vagal terminals to the gastroduodenal area
   B. Stimulation of the sympathetic postganglionic terminals to the gastroduodenal area
   C. Contact with gastric chyme
   D. Peptones   Ref. p. 762

945. IT WOULD BE FALSE TO STATE THAT THE MUCOSA OF THE SMALL INTESTINE:
   A. Regenerates at a very slow speed
   B. Secretes enterokinase, a protein enzyme which has another protein enzyme as its substrate
   C. Secretes more than one liter of neutral isotonic extracellular fluid per day
   D. Is stimulated to secrete at a faster rate by parasympathetic activity
      Ref. pp. 762, 763

946. A SIX MONTH-OLD INFANT WHO HAS LARGE DAILY BOWEL MOVEMENTS WITH A LIQUID CONSISTENCY, DUE TO A CHRONIC INFLAMMATION OF THE LARGE BOWEL, IS LIKELY TO:
   A. Have little difficulty because of the rapid repair of the large bowel
   B. Develop hypercalcemia
   C. Have congestive heart failure and edema as a result of abnormal losses of protein with intestinal contents
   D. Develop hypokalemia
   E. Develop a metabolic alkalosis   Ref. p. 763

947. THE CHEMICAL COMMON DENOMINATOR OF ALL DIGESTIVE PROCESSES TAKING PLACE WITHIN THE LUMEN OF THE GASTROINTESTINAL TRACT CONSISTS OF A(N):
   A. Phosphorylisis
   B. Hydrolysis
   C. Esterification
   D. Condensation
   E. Stereochemical rearrangement   Ref. p. 765

948. THE PROCESS REFERRED TO IN THE PRECEDING QUESTION CONSISTS OF:
   A. The splitting of a complex molecule by a phosphate radical
   B. The splitting of a chemical bond by water
   C. A removal of water from two parent molecules to form an ester
   D. A rotation of the molecule on a tridimensional axis
      Ref. p. 765

SECTION V - GASTROINTESTINAL PHYSIOLOGY

949. THE ACTION OF THE AMYLASE CONTAINED IN SALIVARY SECRETIONS:
   A. Continues to hydrolyze starch in the central portions of the bolus of food in the stomach for a relatively long period of time after swallowing
   B. Accounts for less than 1% of total hydrolysis of disaccharides
   C. Is responsible for 99% of the total hydrolysis of disaccharides
   D. Releases free glucose in the esophagus and stomach
   Ref. p. 766

950. IN THE DIGESTION OF CARBOHYDRATES:
   A. Pancreatic amylase is unable to separate the cellulose cover of uncooked starch granules
   B. Glucose is the final product of the digestive action of pancreatic amylase within the intestinal lumen
   C. The brush border of intestinal cells (small intestine) contains the enzymes which split disaccharides into hexoses
   D. Maltose is split into one molecule of glucose and one molecule of isomaltose
   E. Some disaccharides cannot be split and are absorbed as such into the portal blood
   Ref. p. 766

951. THE FINE EMULSIFICATION OF FATS IN THE WATER MEDIUM IN THE LUMEN OF THE SMALL INTESTINE:
   A. Is not necessarily essential for the quantitative digestion of triglycerides
   B. Is dependent to a major extent on the mixing movements of the small intestine
   C. Could not take place without the polar distribution of fat and water soluble ionized bile salts
   D. Occurs simply because fat globules have a much higher surface tension than water
   Ref. p. 767

952. THE DIGESTION OF FAT:
   A. Could normally occur in the absence of pancreatic lipase if there is normal secretion of gastric lipase
   B. Occurs mainly in the distal portions of the jejunum where bile salts are reabsorbed
   C. Proceeds only on the surface of lipid micelles because lipase is water-soluble
   D. Is not really required as an indispensable process that must precede absorption, as the latter can incorporate into the intestinal lymphatics and capillaries a large bulk of intact triglycerides and phospholipids
   Ref. p. 767

953. THE ENZYMATIC ACTIVITY OF PEPSIN, THE GASTRIC PROTEOLYTIC ENZYME:
   A. Causes the digestion of all proteins with separation of free amino acids in the gastric lumen, if the pH equals 1.00-2.00
   B. Is restricted to a few peptide linkages and leaves intact all others
   C. Leaves collagen intact
   D. May continue its digestive activity in the same medium where chymotrypsin is optimally active
   Ref. p. 768

SECTION V - GASTROINTESTINAL PHYSIOLOGY

954. THE ENZYME(S) RESPONSIBLE FOR THE FINAL RELEASE OF AMINO ACIDS FROM PARENT POLYPEPTIDES:
A. Are amino-peptidases and dipeptidases
B. Is trypsin
C. Is chymotrypsin
D. Is carboxypeptidase Ref. p. 769

955. GLUCOSE IS ABSORBED BY THE SMALL INTESTINE:
A. Through a process which requires the expenditure of energy by the epithelial cells of the intestinal mucosa
B. At a faster rate than any of other known hexoses
C. With either a pyranose or a furanose ring structure
D. By an enzymatic system which is independent of sodium and potassium concentration in the intestinal lumen and only critically dependent on oxygen tension Ref. p. 771

956. THE ABSORPTION OF AMINO ACIDS TAKES PLACE:
A. By diffusion only, as after absorption they rapidly disappear in the circulating portal blood
B. After a delay, because amino acids accumulate in minute micelles in the intestinal lumen, following the digestion of protein molecules
C. Is dependent on an intact sodium transport
D. With preferential transport of D-stereoisomers over that of L-stereoisomers Ref. p. 771

957. THE ABSORPTION OF A NUMBER OF AMINO ACIDS THROUGH THE INTESTINAL LUMEN IS ESSENTIALLY DEPENDENT ON THE LOCAL PRESENCE OF A VITAMIN KNOWN AS:
A. Riboflavin
B. Thiamine
C. Cobalamin
D. Ascorbic acid
E. Pyridoxal phosphate
Ref. p. 771

958. IN THE INTESTINAL ABSORPTION OF FATS:
A. Fatty acids re-esterify directly with glycerol molecules
B. Fatty-acyl-Coenzyme A groups are esterified with glycerophosphate within the epithelial cells of the intestinal mucosa
C. The bulk of fatty acids are freely absorbed into the intestinal lymph without esterification
D. An active transport mechanism establishes a concentration gradient, higher beyond the brush border and lower in the intestinal lumen
Ref. p. 772

959. CHYLOMICRONS TRANSPORTED BY INTESTINAL LYMPH TO THE LIVER ARE COMPOSED OF:
A. Triglycerides only
B. Phospholipids and triglycerides around a protein core
C. Protein around a core of triglycerides and phospholipids
D. A small fraction of triglycerides and a bulk of phospholipids and cholesterol Ref. p. 772

SECTION V - GASTROINTESTINAL PHYSIOLOGY

960. ALL OF THE FOLLOWING FACTORS INCREASE AND PROMOTE FAT ABSORPTION, EXCEPT:
   A. Bile salts in the intestinal lumen
   B. Active transport mechanism for fatty acids
   C. Fatty-acyl-Coenzyme A esterification with glycerophosphate within epithelial cells
   D. Intestinal circulation of lymph
   E. Intestinal circulation of portal blood   Ref. p. 772

961. IN RELATION TO THE INTESTINAL ABSORPTION OF SODIUM, IT SHOULD BE REMEMBERED THAT:
   A. It occurs through an active transport mechanism which operates in the basal and lateral parts of the epithelial cells of the intestinal mucosa
   B. The bulk of this absorption takes place in the ileum and colon
   C. A large concentration gradient of sodium (intestinal lumen/portal blood) is usually established because of the predominant influx versus efflux
   D. Quantitatively it is second to calcium absorption, which is aided by the hormonal influence of parathyroid hormone and vitamin D
        Ref. p. 773

962. THE LARGEST COMPONENT (ON A WEIGHT BASIS) OF HUMAN FECES IS:
   A. Water
   B. Dead bacteria
   C. Live bacteria
   D. Protein
   E. Fat
        Ref. p. 774

963. MEGAESOPHAGUS IS FREQUENTLY ASSOCIATED WITH:
   A. Achalasia
   B. Myasthenia gravis
   C. Botulism
   D. Scleroderma
   E. Dystonia
        Ref. p. 775

964. A COMPLETE GASTRECTOMY WILL USUALLY:
   A. Prevent protein digestion from taking place, as pepsin and hydrochloric acid are essential for the bulk of protein digestion
   B. Cause malabsorption of vitamin B_{12}
   C. Suppress only half of the source of intrinsic factor
   D. Make the person constantly hungry because the reservoir function of the stomach is essential to establish a normal feeding cycle
        Ref. p. 776

965. THE REASON WHY A PATIENT WITH PEPTIC ULCER SHOULD BE FED AT SHORT INTERVALS AND NOT 3 TIMES EVERY DAY, CONSISTS IN THE FACT THAT, WITH FREQUENT FEEDINGS:
   A. Pyloric emptying is more efficient
   B. Vagal activity is less likely
   C. The gastrin mechanism does not operate as well
   D. The anxiety caused by hunger is eliminated
   E. Food in the stomach neutralizes the acid
        Ref. p. 777

SECTION V - GASTROINTESTINAL PHYSIOLOGY

966. THE RESTRICTION OF ALCOHOL AND MEAT EXTRACTS TO PATIENTS UNDERGOING DIETARY TREATMENT FOR A PEPTIC ULCER IS BASED ON THE NOTION THAT THESE SUBSTANCES:
    A. Accelerate gastric emptying and expose the duodenal mucosa to more frequent acid stimuli
    B. Suppress mucus secretion and eliminate a natural defensive mucosal barrier
    C. Inhibit the secretion of enterogastrone
    D. Stimulate the secretion of gastrin
    E. Are potent histaminergics   Ref. p. 777

967. THE ADDITION OF FAT-RICH CREAM TO THE DIET OF THE PATIENT WITH AN ACUTE PEPTIC ULCER IS LOGICAL BECAUSE FOOD HIGH IN LIPID CONTENT:
    A. Inhibits gastricsin in the stomach
    B. Inhibits gastrin secretion in the pyloric antrum
    C. Slows gastric emptying
    D. Neutralizes hydrochloric acid as a buffer
    E. Accelerates mucosal repair   Ref. p. 777

968. SUBTOTAL GASTRECTOMY CONTINUES TO BE AN UNAVOIDABLE FORM OF TREATMENT IN MANY PROTRACTED CASES OF PEPTIC (DUODENAL) ULCER, BECAUSE THE REMOVAL OF A LARGE PART OF THE STOMACH REMOVES:
    A. The source of pain
    B. The source of excessive volumes of acid gastric juice rich in pepsin
    C. Most of the postganglionic parasympathetic fibers, which are ultimately the cause of the disease
    D. The pacemaker smooth muscle fibers responsible for excessive gastric contractions which tend to prevent healing
                                            Ref. p. 777

969. MOST PATIENTS WITH CHRONIC PEPTIC ULCER:
    A. Have been demonstrated to secrete large volumes of acid gastric secretions rich in pepsin during the night, while asleep
    B. Are achlorhydrics
    C. Are very calm individuals with no apparent stress acting on them
    D. Will not develop recurrent ulcers in the opening that communicates the gastric stump (of a gastrectomy) with the portion of the jejunum to which the latter has been anastomosed
                                            Ref. p. 777

970. IN CASES OF SPRUE THE FUNDAMENTAL CAUSE OF THE WEIGHT LOSS IS AN:
    A. Impaired digestion of fats
    B. Impaired digestion of proteins
    C. Impaired absorption of digested fat and proteins
    D. Accelerated transit of intestinal contents
    E. Excessively rapid gastric emptying which prevents pancreatic secretions to reach their full concentration and volume
                                            Ref. p. 778

SECTION V - GASTROINTESTINAL PHYSIOLOGY

971. IN WHAT IS KNOWN AS MEGACOLON:
A. Peristaltic movements are greatly increased
B. Diarrhea is quite common
C. A large proportion of water absorption may occur through the large bowel
D. The Auerbach plexus is commonly atrophic
Ref. p. 779

972. IN THE ACT OF VOMITING THE REVERSE PERISTALSIS SET IN MOTION IN THE STOMACH ITSELF IS CONCOMITANT WITH:
A. Closure of esophago-gastric junction
B. Opening of the crico-esophageal sphincter
C. Elevation of the diaphragm
D. Stress relaxation of the abdominal wall
E. Lowering of the hyoid bone and the larynx
Ref. p. 780

973. A LIKELY REASON WHY A TO AND FRO MOTION, SUCH AS THAT ENCOUNTERED DURING AIRPLANE FLIGHTS OR IN AN AUTOMOBILE RIDE OVER A BUMPY ROAD, TENDS TO CAUSE NAUSEA AND VOMITING, IS BECAUSE DURING SUCH MOTIONS:
A. Much air is swallowed and the stomach becomes distended
B. The stomach is more likely to develop reverse peristalsis
C. The cerebral cortex can no longer inhibit an intrinsic tendency of the brain stem to cause vomiting
D. Vestibular reflexes eventually excite a "chemoreceptor trigger zone" in the medulla
E. The cerebellum is strongly inhibited   Ref. p. 781

974. THE SO-CALLED CHEMORECPTOR TRIGGER ZONE:
A. Is synonymous with the vomiting center
B. Is located in the cerebral peduncles, ventral to the Aqueduct of Sylvius
C. Is sensitive to the action of morphine
D. May be destroyed experimentally; the ablation of both chemoreceptor trigger zones abolishes all forms of vomiting
Ref. p. 781

975. A PYLORIC OBSTRUCTION IN AN INFANT WHICH RESULTS IN A PROLONGED BOUT OF SEVERE VOMITING IS LIKELY TO CAUSE A PRIMARY:
A. Metabolic acidosis
B. Metabolic alkalosis
C. Respiratory acidosis
D. Respiratory alkalosis
Ref. p. 782

976. AN OBSTRUCTION OF THE SMALL BOWEL LEADING TO SEVERE VOMITING WITH A PREPONDERANT LOSS OF DUODENAL CONTENTS IS LIKELY TO CAUSE A PRIMARY:
A. Metabolic acidosis
B. Metabolic alkalosis
C. Respiratory acidosis
D. Respiratory alkalosis
Ref. p. 782

SECTION V - GASTROINTESTINAL PHYSIOLOGY

977. A TWISTING OF THE LARGE BOWEL MAY CLOSE THE MESENTERIC VESSELS BY ROTATION OVER THEIR LONGER AXIS. THE LARGE BOWEL BECOMES DILATED. IN THIS CASE A PATIENT WHO DOES NOT RECEIVE MEDICAL ATTENTION (I.E. INTRAVENOUS FLUID THERAPY) IS LIKELY TO DEVELOP:
   A. A low hematocrit
   B. A low plasma volume
   C. Collateral circulation to the infarcted bowel
   D. Relief of the intestinal distension by diffusion of gases through the intestinal wall   Ref. p. 782

SECTION VI - METABOLISM, TEMPERATURE REGULATION AND HEPATIC PHYSIOLOGY

FOR EACH OF THE FOLLOWING MULTIPLE CHOICE QUESTIONS, SELECT THE ONE MOST APPROPRIATE ANSWER:

978. THE CONVERSION OF ADENOSINE-TRIPHOSPHATE TO ADENOSINE-DIPHOSPHATE IN LIVING MAMMALIAN CELLS AT NORMAL BODY TEMPERATURE IS BELIEVED TO CAUSE THE RELEASE OF ABOUT:
   A. 1,000 calories
   B. 2,000 calories
   C. 5,000 calories
   D. 8,000 calories
   E. 10,000 calories
   Ref. p. 787

979. WITHIN MAMMALIAN CELLS ADENOSINE-TRIPHOSPHATE IS KNOWN TO BE DISTRIBUTED:
   A. Only in the nuclei
   B. Only in the endoplasmic reticulum
   C. Only in mitochondria
   D. Only in the cell membrane
   E. In all cellular structures
   Ref. p. 787

980. GLUCOSE IS TRANSFERRED FROM INTERSTITIAL SPACES INTO THE CYTOPLASM OF CELLS AFTER TRAVERSING THE CELLULAR MEMBRANE. IN THE CASE OF ADIPOSE TISSUE CELLS AND SKELETAL MUSCLE FIBERS, AS WELL AS IN FEW OTHER CELLS, THIS PASSAGE PROBABLY OCCURS BY:
   A. Facilitated diffusion
   B. Active transport
   C. Simple diffusion
   D. Reverse transport
   E. Pinocytosis
   Ref. p. 788

981. THE MECHANISM WHICH MEDIATES GLUCOSE TRANSFER THROUGH CELLULAR MEMBRANES OF ADIPOSE TISSUE CELLS AND MUSCLE CELLS:
   A. Creates a concentration gradient with higher concentration of glucose inside the cell than outside
   B. Operates without concentration gradients being created at any time
   C. Operates because there is a higher concentration of glucose out of the cell and a lower concentration inside the cell
   D. Does not follow the rules of enzyme kinetics
   E. Does not change its basic rate if the cells are hypoxic
   Ref. p. 788

982. THE MECHANISM WHICH FACILITATES GLUCOSE TRANSFER INTO THE CELLULAR CYTOPLASM OF SKELETAL MUSCLE FIBERS IS HORMONALLY PROMOTED BY:
   A. Epinephrine
   B. Norepinephrine
   C. Glucagon
   D. Insulin
   E. Vasopressin
   Ref. p. 788

983. TO A VERY LIMITED EXTENT DISACCHARIDES CAN PASS THE CELLULAR MEMBRANES OF:
   A. Renal tubular cells
   B. Myocardial cells
   C. Skeletal muscle fibers
   D. Liver cells
   E. Epithelial cells of the intestinal mucosa
   Ref. p. 788

SECTION VI - METABOLISM, TEMPERATURE REGULATION AND HEPATIC PHYSIOLOGY

984. GLUCAGON INCREASES GLYCOGEN BREAKDOWN IN LIVER CELLS BY INCREASED:
    A. Synthesis of glucose from amino acids
    B. Synthesis of glucose from pyruvate
    C. Synthesis of glycogen through activation of glycogen synthetase
    D. Synthesis of cyclic AMP
    E. Activity of glucose-6-phosphatase   Ref. p. 790

985. ADENOSINTRIPHOSPHATE (ATP) IS REQUIRED AT TWO STEPS IN THE CHAIN REACTION OF THE GLYCOLYTIC CYCLE. ONE OF THESE STEPS WHICH UTILIZES AVAILABLE ATP OCCURS WHEN:
    A. Glucose is phosphorylated to form glucose-6-phosphate
    B. Glucose-6-phosphate is converted into fructose-6-phosphate
    C. Fructose 1, 6-diphosphate is split by aldolase to form dihydroxy-acetone and glyceraldehyde-3-phosphate
    D. Glyceraldehyde-3-phosphate is oxidized to form diphosphoglyceric acid
    E. Phosphoenolpyruvic is formed from 2-phosphoglyceric acid
    Ref. p. 791

986. FOUR MOLES OF ATP CAN BE FORMED FROM 1 MOL OF GLUCOSE DURING THE GLYCOLYTIC CYCLE. TWO OF THESE ARE SYNTHESIZED WHEN:
    A. 2 mols of phosphoenolpyruvic acid are converted into pyruvic acid
    B. Glucose-6 phosphate is converted into fructose-6-phosphate
    C. 2 mols of 3-phosphglyceric acid are transformed into 2 mols of 2-phosphoglyceric acid
    D. Aldolase splits fructose 1, 6-diphosphate
    E. Dihydroxyacetone phosphate is transformed into glyceraldehyde-3-phosphate   Ref. p. 791

987. COENZYME A, WHICH PLAYS AN INDISPENSABLE ROLE IN LIPID AND GLUCOSE METABOLISM, HAS AMONG ITS COMPONENTS:
    A. Pyridoxine
    B. Thiamine
    C. Riboflavin
    D. Pantothenic acid
    E. Ascorbic acid
    Ref. p. 791

988. THE FORMATION OF 2 MOLS OF ACETYL-COENZYME A FROM 2 MOLS OF PYRUVIC ACID:
    A. Gives rise directly to 2 mols of ATP
    B. Uses up 2 mols of ATP
    C. Incorporates CO_2 into the structure of acetyl Coenzyme A
    D. Releases 4 hydrogen atoms   Ref. p. 792

989. CITRIC ACID RESULTS FROM THE CHEMICAL COMBINATION OF ACETYL CoA WITH:
    A. Cis-aconitic acid
    B. Iso-citric acid
    C. Oxalosuccinic acid
    D. Fumaric acid
    E. Oxaloacetic acid   Ref. p. 792

SECTION VI - METABOLISM, TEMPERATURE REGULATION AND HEPATIC PHYSIOLOGY

990. THE ABOVE DESCRIBED REACTION:
    A. Utilizes ATP
    B. Forms ATP
    C. Releases hydrogen atoms
    D. Releases water
    E. Incorporates water
    Ref. p. 792

991. THE COMPLETE UTILIZATION OF 2 MOLS OF ACETYL COENZYME A IN THE TRICARBOXYLIC ACID CYCLE CONTRIBUTES TO THE FORMATION OF ATP, FUNDAMENTALLY THROUGH THE:
    A. Release of CO_2 and its subsequent oxidation
    B. Release of hydrogen atoms and their ultimate oxidation
    C. Stepwise use of water
    D. Use of TPN as the major cofactor   Ref. p. 792

992. THE LARGE PROTEIN ENZYMES WHICH CATALYZE THE TRANSFER OF HYDROGEN ATOMS TO DPN ARE KNOWN AS:
    A. DPN-ases
    B. Cytochrome oxidases
    C. Dehydrogenases
    D. Oxidases   Ref. p. 793

993. THE RELEASE OF HYDROGEN ATOMS FROM VARIOUS MOLECULES IN THE TRICARBOXYLIC ACID CYCLE CONSTITUTES A KEY REACTION LEADING TO THE STORAGE OF PART OF THE ENERGY (RELEASED) IN THE FORM OF ATP. THE FOUR PARENT MOLECULES CAPABLE OF RELEASING 2 HYDROGEN ATOMS AS THEY UNDERGO CHEMICAL TRANSFORMATION IN THE TCA CYCLE ARE:
    A. Isocitric acid, alpha ketoglutaric acid, succinic acid and malic acid
    B. Oxaloacetic acid, fumaric acid, citric acid and cis-aconitic acid
    C. Oxalosuccinic acid, fumaric acid, oxaloacetic acid and citric acid
    D. Oxaloacetic acid, isocitric acid, oxalosuccinic acid and citric acid
    Ref. p. 792

994. OXIDATIVE PHOSPHORYLATION IS A PROCESS IN WHICH:
    A. Adenosin-diphosphate is converted into adenosin-triphosphate while alpha ketoglutaric acid is transformed into succinic acid
    B. Electrons are removed from hydrogen atoms and transferred to oxygen with release of an excess energy used to convert ADP into ATP
    C. A glycogen molecule is phosphorylated to form glucose-1-phosphate with the participation of oxygen
    D. Glucose is phosphorylated to form glucose-6-phosphate with the participation of oxidative enzymes   Ref. p. 793

995. AS A RESULT OF THE SERIES OF REACTIONS OF THE TRICARBOXYLIC ACID CYCLE TWO MOLECULES OF ATP ARE FORMED DIRECTLY FOR EACH GLUCOSE MOLECULE UNDERGOING OXIDATION. IN COMPARISON WITH THIS SMALL YIELD, THE SUBSEQUENT OXIDATION OF HYDROGEN ATOMS RELEASED AS PART OF THE SERIES OF REACTIONS OF THE TRICARBOXYLIC ACID CYCLE GIVES, FOR EACH MOLECULE OF GLUCOSE, A TOTAL YIELD OF:
    A. 14 molecules of ATP per 1 molecule of glucose
    B. 24 molecules of ATP per 1 molecule of glucose
    C. 34 molecules of ATP per 1 molecule of glucose
    D. 44 molecules of ATP per 1 molecule of glucose
    E. 84 molecules of ATP per 1 molecule of glucose
    Ref. p. 795

SECTION VI - METABOLISM, TEMPERATURE REGULATION
AND HEPATIC PHYSIOLOGY

996. COMPARED TO 304 KILOCALORIES THAT MAY BE STORED AS ATP BY TOTAL OXIDATION OF ONE MOL OF GLUCOSE, THE USEFUL ENERGY THAT MAY BE STORED DURING ANAEROBIC GLYCOLYSIS AMOUNTS ONLY TO:
   A. 1 Kilocalorie
   B. 2 Kilocalories
   C. 8 Kilocalories
   D. 16 Kilocalories
   E. 32 Kilocalories
   Ref. p. 795

997. A BYPRODUCT OF ANAEROBIC GLYCOLYSIS IS:
   A. Acetylcoenzyme A
   B. Thiamine
   C. Citric acid
   D. Lactic acid
   E. Fumaric acid
   Ref. p. 795

998. A MAJOR METABOLIC USE IS MADE OF TPNH (TRIPHOSPHOPYRIDINE NUCLEOTIDE, IN REDUCED FORM) IN THE SEQUENCE OF EVENTS WHICH:
   A. Culminate in gluconeogenesis
   B. Form glycogen
   C. End in protein synthesis
   D. Cause fat catabolism
   E. Induce fatty acid synthesis
   Ref. p. 797

999. IN AN INDIRECT OR DIRECT FASHION GLUCONEOGENESIS IS ENHANCED BY THE INCREASED SECRETION OF:
   A. Epinephrine
   B. Norepinephrine
   C. Glucagon
   D. Aldosterone
   E. ACTH
   Ref. p. 797

1000. IT IS BELIEVED THAT THE RATE OF GLYCOLYSIS MAY BE MARKEDLY REDUCED:
   A. If oxygen is not available
   B. Under anaerobic conditions because lactic acid cannot diffuse out of the cells
   C. If ATP is scarce
   D. If ADP is scarce
   E. If enzymes of the tricarboxylic acid cycle are competitively inhibited
   Ref. p. 795

1001. THE SO-CALLED PENTOSE SHUNT (OR PHOSPHOGLUCONATE PATHWAY) FOR GLUCOSE METABOLISM:
   A. Is a sequence of metabolic reactions which preferentially occurs when there is cellular anaerobiosis
   B. Does not originate any ATP but consumes available ATP
   C. Releases hydrogen atoms which are oxidized with the aid of a coenzyme that has an additional phosphate group over those present in DPN
   D. Is the predominant form of oxidation of glucose in skeletal muscle
   E. Is the least important means of oxidation of glucose in adipose tissue cells
   Ref. p. 797

1002. MOST OF BLOOD GALACTOSE IS UTILIZED WITHIN CELLS TO FORM:
   A. Fructose
   B. Amino acids
   C. Lactose
   D. Glucose
   E. Proteins found in milk
   Ref. p. 797

SECTION VI - METABOLISM, TEMPERATURE REGULATION AND HEPATIC PHYSIOLOGY

1003. AMONG THE VARIOUS LIPID GROUPS PRESENT IN MAMMALIAN TISSUES, THE MOST ABUNDANT SOURCE OF POTENTIAL CHEMICAL ENERGY IS IN THE FORM OF:
A. Triglycerides
B. Fatty acids in free from
C. Cholesterol
D. Phospholipids
E. Cholesterol esters
Ref. p. 799

1004. THE FATTY ACID WHICH HAS 18 CARBONS AND ONE DOUBLE BOND INTERPOSED IN THE LONG ALIPHATIC CARBON CHAIN (CIS 9-OCTADECENOIC ACID) IS KNOWN AS:
A. Palmitic acid
B. Stearic acid
C. Oleic acid
D. Palmitoleic acid
E. Linoleic acid
Ref. p. 799

1005. DURING PERIODS WHEN BLOOD GLUCOSE LEVELS ARE LOW, THE SYNTHESIS OF NEW GLUCOSE MOLECULES TAKES PLACE AT A RELATIVELY RAPID RATE PREDOMINANTLY IN THE:
A. Brain
B. Heart
C. Liver
D. Skeletal muscle
E. Adipose tissue
Ref. p. 797

1006. A LARGE PORTION OF THE CARBON CHAINS WHICH EVENTUALLY FORM GLUCOSE ARE DERIVED FROM:
A. Citric acid
B. Carbonic acid
C. Uric acid
D. Acetoacetic acid
E. Amino acids
Ref. p. 797

1007. THE HORMONE ENDOWED WITH A PREDOMINANT CAPACITY TO CATABOLYZE PROTEINS UTILIZED IN THE GLUCONEOGENETIC PROCESS IS:
A. Vasopressin
B. Aldosterone
C. Progesterone
D. Pregnenolone
E. Cortisol
Ref. p. 797

1008. IT IS TRUE THAT:
A. Betalipoproteins have a higher concentration of cholesterol than other lipoproteins
B. Lipoproteins with the highest diameters have the highest density within the group of blood lipoproteins
C. Chylomicrons owe their stability in plasma to the fact that they have the largest percentage of protein of all lipoproteins
D. The highest density shown by lipoproteins equals about 1.4
Ref. p. 800

1009. FATTY ACIDS PRESENT IN MAMMALIAN BLOOD:
A. Are transported in loose combination with phospholipids
B. Have a very short half-life in blood after they are released from storage forms or after they are absorbed from the intestinal lumen
C. Are for the most part composed of structural chains not longer than 10 carbons
D. Are usually not esterified with cholesterol until they reach the intimal layer of arteries
Ref. p. 800

SECTION VI - METABOLISM, TEMPERATURE REGULATION AND HEPATIC PHYSIOLOGY

1010. POSTPRANDIAL PLASMA IS OFTEN TIMES LACTESCENT BECAUSE OF THE HIGH CONCENTRATION OF LARGE SIZED TRIGLYCERIDE-RICH CHYLOMICRONS. AN ACCELERATED REMOVAL OF THESE TRIGLYCERIDES FROM CIRCULATING BLOOD CAN BE OBTAINED BY INJECTIONS OF:
   A. Insulin
   B. Cholesterol
   C. Epinephrine
   D. Heparin
   E. Protamine
   Ref. p. 800

1011. AMONG LIPOPROTEINS, THE HIGHEST S_f COEFFICIENT VALUES ARE THOSE OF:
   A. Chylomicrons
   B. Beta-lipoproteins
   C. Prebeta-lipoproteins
   D. Alpha lipoproteins
   Ref. p. 801

1012. LIPOPROTEINS CIRCULATING IN BLOOD DURING FASTING PERIODS:
   A. Originate for the most part in the gastrointestinal mucosa
   B. Carry a large concentration of high S_f components
   C. Transport a significant amount of fat newly synthesized from carbohydrates
   D. Are predominantly of the high molecular type
   Ref. p. 801

1013. ADIPOSE TISSUE CELLS:
   A. Contain triglyceride that turns over about once every six months
   B. Contain triglyceride which is mainly composed by short chain fatty acids
   C. Contain more cholesterol than triglyceride
   D. Cannot convert glucose into triglyceride molecules as this transformation is restricted to the liver cells
   E. Contain triglyceride in a liquid state
   Ref. p. 802

1014. THE INITIAL STEP IN THE CATABOLIC OXIDATION OF FATTY ACIDS, KNOWN AS BETA OXIDATION, IS THE COMBINATION OF A FATTY ACID WITH:
   A. DPN (diphospho-pyridine nucleotide)
   B. TPN (triphospho-pyridine nucleotide)
   C. ADP (adenosine-diphosphate)
   D. CoA (Coenzyme A)
   E. Glycerol
   Ref. p. 802

1015. ADENOSINE-TRIPHOSPHATE IS USED IN THE CATABOLIC OXIDATION OF FATTY ACIDS (BETA OXIDATION) IN THE STEP WHERE:
   A. A fatty acyl-Coenzyme A loses 2 hydrogen atoms from 2 contiguous carbons (the alpha and beta carbons) to flavoproteins
   B. A hydroxy-acyl-Coenzyme A loses 2 hydrogen atoms to DPN
   C. Water is incorporated in the oxidized acyl-Coenzyme A between the alpha and the beta carbon
   D. Acetyl-Coenzyme A combines with oxyaloacetate to begin oxidation in the tricarboxylic acid cycle
   Ref. p. 803

SECTION VI - METABOLISM, TEMPERATURE REGULATION AND HEPATIC PHYSIOLOGY

1016. THE TERMINAL CARBOXYLIC GROUP IS ESTERIFIED TO COENZYME A IN THE ACYL-CoA RADICALS FORMED AT THE START OF THE SO-CALLED BETA OXIDATION OF FATTY ACIDS. IF THE CARBON NEXT TO THE CARBOXYLIC GROUP IS NAMED ALPHA AND THE SUBSEQUENT CARBONS ARE CALLED BETA, GAMMA, EPSILON, THE BREAK BETWEEN CARBONS IN BETA OXIDATION OCCURS BETWEEN THE:
 A. Carboxylic carbon and the alpha carbon
 B. Alpha carbon and the beta carbon
 C. Beta carbon and the gamma carbon
 D. Gamma carbon and the epsilon carbon
 Ref. p. 803

1017. IN THE BETA OXIDATION OF FATTY ACIDS, ATP SYNTHESIS:
 A. Occurs only after acetyl Coenzyme A enters the tricarboxylic acid cycle but not as a result of reactions occurring during the formation of acetyl Coenzyme A
 B. Takes place after hydrogen atoms are removed from fatty acids in the process that leads to the formation of acetyl Coenzyme A
 C. Cannot occur because the whole reaction uses up available ATP
 D. Is equal to ATP utilization leading to a zero net balance of ATP molecules
 Ref. p. 803

1018. BECAUSE OF THE LARGE NUMBER OF PAIRED CARBONS IN LONG CHAIN FATTY ACIDS (SO ABUNDANT IN FAT DEPOTS) THE POTENTIAL YIELD OF ATP RESULTING FROM THESE PAIRED CARBONS ENTERING THE TRICARBOXYLIC ACID CYCLE AS ACETYL COENZYME A EQUALS ABOUT:
 A. 16 ATP molecules for each pair of carbons
 B. 26 ATP molecules for each pair of carbons
 C. 36 ATP molecules for each pair of carbons
 D. 46 ATP molecules for each pair of carbons
 Ref. p. 803

1019. BLOOD TRANSPORTS THE RESULT OF HEPATIC BETA OXIDATION OF FATTY ACIDS IN THE PREDOMINANT MOLECULAR FORM OF:
 A. Acetyl Coenzyme A
 B. Acetoacetic acid
 C. Beta-hydroxybutyric acid
 D. Acetone
 E. Acyl-Coenzyme A
 Ref. p. 804

1020. A SIGNIFICANT RISE OF ACETOACETIC ACID DURING PROTRACTED STARVATION IS ASSOCIATED IN THE NORMAL HUMAN WITH A(N):
 A. High insulin titer
 B. Low rate of cortisol secretion
 C. Low concentration of sodium bicarbonate
 D. Low rate of ACTH secretion
 E. Accelerated anaerobic glycolytic cycle in skeletal muscle
 Ref. p. 804

SECTION VI - METABOLISM, TEMPERATURE REGULATION AND HEPATIC PHYSIOLOGY

1021. THE MAIN SOURCE OF ACETYL COENZYME A USED IN THE SYNTHESIS OF FATTY ACIDS IS:
A. The acetoacetate molecules that originate in the liver
B. Oxaloacetate in the tricarboxylic acid cycle
C. Citric acid
D. Pyruvic acid
E. Triglycerides   Ref. p. 805

1022. THE MAIN SOURCE OF TPNH UTILIZED DURING FATTY ACID SYNTHESIS IS THE:
A. Tricarboxylic acid cycle
B. Embden Meyerhof cycle (glycolysis)
C. Catabolism of triglycerides caused by tissue lipase
D. Pentose shunt
E. Transamination process taking place in the liver
           Ref. p. 805

1023. TRIGLYCERIDE SYNTHESIS REQUIRES THAT THERE BE AN INCREASED RATE OF SECRETION OF:
A. Insulin
B. Growth hormone
C. Corticotrophin (ACTH)
D. Epinephrine   Ref. p. 805

1024. AN ESSENTIAL INTERMEDIATE COMPONENT IN THE SYNTHESIS OF TRIGLYCERIDES IS THE:
A. Glycerol molecule
B. Glyceraldehyde molecule
C. Alpha-glycerophosphate molecule
D. Dihydroxyacetone molecule
E. Acetoacetate
           Ref. p. 806

1025. WEIGHT LOSS DUE TO INSUFFICIENCY IN THE SYNTHESIS OF TRIGLYCERIDES IS AN INEVITABLE CONSEQUENCE WHEN THERE IS HORMONAL DEFICIENCY OF:
A. Growth hormone
B. Epinephrine
C. Thyroid hormone
D. Insulin
E. Glucagon
           Ref. p. 806

1026. ASIDE FROM NORMAL AVAILABILITY OF TISSUE OXYGEN, A PRIME DETERMINANT OF THE RATE OF OXIDATION OF SUBSTRATE IS THE NORMAL CONCENTRATION OR SYNTHESIS OF:
A. Acetylcoenzyme A
B. Pyruvic acid
C. Citrate
D. Glucose
E. Adenosin-diphosphate (ADP)
           Ref. p. 807

1027. IF A STARVING ADULT MAN WERE TO USE UP HIS GLYCOGEN STORES AFTER THE FIRST 24 HOURS (ONE DAY) OF FASTING, THE FAT DEPOTS COULD THEORETICALLY PROVIDE ENERGY FOR THE NEXT:
A. Two days
B. Eight days
C. Fifteen days
D. Thirty days
E. Sixty days   Ref. p. 806

SECTION VI - METABOLISM, TEMPERATURE REGULATION AND HEPATIC PHYSIOLOGY

1028. INHIBITION OF UTILIZATION OR CATABOLISM OF TRIGLYCERIDES OCCURS:
   A. During starvation
   B. If insulin is absent, as in diabetes mellitus
   C. After a glucose rich diet in a person who secretes insulin at a normal rate
   D. When growth hormone is secreted in response to hypoglycemia
   E. In prolonged physical exercise   Ref. p. 807

1029. AN ACTIVE TRICARBOXYLIC ACID CYCLE FAVORS A CONCOMITANT:
   A. Increase in fatty acid beta-oxidation
   B. Accumulation of acetyl-Coenzyme A
   C. Carboxylation of acetyl-Coenzyme A to form malonyl-Coenzyme A
   D. Decrease in glycolysis   Ref. p. 807

1030. THE ABOVE REACTION RESULTS IN:
   A. Disappearance by catabolism of a large number of fatty acid molecules from fat depots
   B. Acidosis
   C. Fatty acid and triglyceride synthesis
   D. Accumulation of glucose 6 phosphate
                                    Ref. p. 807

1031. THE BULK OF BLOOD PHOSPHOLIPIDS ARE FORMED IN:
   A. Red blood cells   D. Liver cells
   B. White blood cells (leukocytes)   E. Adipose tissue cells
   C. Platelets   Ref. p. 808

1032. THE MYELIN LAYER WHICH SURROUNDS LARGE, FAST CONDUCTING AXONS, IS COMPOSED MAINLY OF:
   A. Cholesterol   D. Phosphatidylcholine
   B. Triglycerides   E. Sphingomyelin
   C. Lecithin   Ref. p. 808

1033. A UNIQUE FEATURE OF CHOLESTEROL IS THAT:
   A. It always is esterified with fatty acids before it can be absorbed by the small intestine
   B. Most of endogenous cholesterol is formed in the fat depot and is transported for degradation to the liver by lipoproteins
   C. It is the only lipid which is relatively water soluble
   D. It does not need to be digested in order to undergo intestinal absorption   Ref. p. 808

1034. ONE POSSIBLE REASON WHY A TRIGLYCERIDE RICH DIET, CONTAINING SATURATED FATTY ACIDS, TENDS TO INCREASE THE BLOOD CHOLESTEROL CONCENTRATION IS BECAUSE:
   A. The glycerol moiety of triglycerides is utilized to form the cholesterol ring structure
   B. The fatty acids from triglycerides promote cholesterol synthesis by increasing the concentration of acetyl Coenzyme A which is utilized directly in the synthesis of new cholesterol
   C. Cholesterol degradation is inhibited by triglycerides in the intestinal lumen
   D. Bile salt reabsorption is inhibited by a triglyceride-rich diet
                                    Ref. p. 809

SECTION VI - METABOLISM, TEMPERATURE REGULATION AND HEPATIC PHYSIOLOGY

1035. AN INCREASED BLOOD CHOLESTEROL USUALLY FOLLOWS A PROLONGED REDUCTION IN THE SECRETION OF:
A. Growth hormone
B. Corticotrophin
C. Glucagon
D. Thyroxine
E. Cortisol
Ref. p. 809

1036. A HORMONE PRESUMED TO BE CAPABLE OF LOWERING BLOOD CHOLESTEROL IN HUMANS IS:
A. Parathyroid hormone
B. Estrogens
C. Androgens
D. Cortisol
E. Glucagon
Ref. p. 809

1037. CHOLIC ACID AND ITS CONGENERS ARE PRODUCTS OF CHOLESTEROL METABOLISM. THEY ARE:
A. Purely excretory products without any useful function once they are formed in the liver
B. A minimal fraction of cholesterol catabolism
C. Strong acids that constantly threaten the acid base balance of the organism
D. Conjugated to taurine and glycine and play an essential role in fat digestion and absorption   Ref. p. 809

1038. THE EXPERIMENTAL INDUCTION OF ATHEROSCLEROSIS IN DOGS IS NOT AS EASY AS IN RABBITS. IN ORDER TO CAUSE ATHEROMATA IN DOGS THEY MUST RECEIVE CHOLESTEROL RICH DIETS, PLUS:
A. A drug that inhibits insulin release
B. Thyroid hormones
C. A drug that inhibits androgen release
D. Estrogens
E. A drug that inhibits thyroxine synthesis
Ref. p. 810

1039. THE HEMODYNAMIC FACTOR WHICH IS BEST CORRELATED WITH THE PREMATURE DEVELOPMENT OF ATHEROSCLEROSIS IS:
A. A high viscosity of blood due to a high hematocrit
B. A low viscosity of blood due to a low hematocrit
C. A relatively low mean linear blood velocity during the rapid ejection phase of cardiac systole
D. High systolic and low diastolic blood pressure
E. High mean arterial blood pressure   Ref. p. 811

1040. AMINO ACID CONCENTRATION IN BLOOD:
A. Has a relatively small increment after protein meals
B. Varies over a large range in a 24 hour period which depends on the rate of cellular uptake of amino acids at different times of the day
C. Reaches very high concentrations after a protein rich meal
D. Varies from hour to hour mainly due to fluctuations in renal tubular reabsorption of amino acids   Ref. p. 815

SECTION VI - METABOLISM, TEMPERATURE REGULATION   195
AND HEPATIC PHYSIOLOGY

1041. THE KIDNEY:
A. Filters most of the circulating plasmatic amino acids and reabsorbs them through an active transport process
B. Cannot filter amino acids through the glomerular membrane because of the relatively large size of these molecules
C. Has no T_m for filtered amino acids
D. Reabsorbs amino acids in the epithelial cells of the collecting ducts
Ref. p. 815

1042. THE BULK OF AMINO ACIDS PASS FROM PLASMA INTO THE MAJORITY OF CELLS BY A PROCESS WHICH INVOLVES:
A. Diffusion
B. Facilitated diffusion
C. Active transport
D. Pinocytosis   Ref. p. 815

1043. LYMPHATIC TISSUE IN LYMPH NODES AND THE SPLEEN CONTRIBUTES TO THE FORMATION OF:
A. Fibrinogen
B. Albumin
C. Gamma-globulin
D. Prealbumin   Ref. p. 816

1044. THE EXCESSIVELY HIGH LOSSES OF PROTEIN THROUGH AN ABNORMAL KIDNEY ARE COMPENSATED IN PART BY A POTENTIAL PROTEIN SYNTHETIC CAPACITY IN THE LIVER WHICH CAN REPLENISH THE PROTEIN CONTENT OF:
A. 10 ml plasma/day   D. 3000 ml plasma/day
B. 100 ml plasma/day   E. 5000 ml plasma/day
C. 1000 ml plasma/day   Ref. p. 816

1045. PARENTERAL THERAPY OF PROTEIN DEFICIENCY IS:
A. Physiologically sound, as plasma and tissue proteins are rapidly exchanged
B. Useless, because most of injectable protein causes severe anaphylaxis
C. Useless, because exogenous amino acids cannot be utilized by the liver to synthesize new protein
D. Not advisable for more than 24 hours because of rapid renal clearance of exogenous amino acids   Ref. p. 816

1046. ALL OF THE FOLLOWING ARE ESSENTIAL AMINO ACIDS, EXCEPT:
A. Glutamic acid   D. Leucine
B. Phenylalanine   E. Methionine
C. Tryptophan   Ref. p. 813

1047. ONE OF THE FOLLOWING IS AN ESSENTIAL AMINO ACID WHILE THE REMAINING FOUR ARE NON ESSENTIAL. MARK THE ESSENTIAL AMINO ACID:
A. Glycine
B. Alanine
C. Arginine
D. Cysteine
E. Tyrosine   Ref. p. 813

SECTION VI - METABOLISM, TEMPERATURE REGULATION AND HEPATIC PHYSIOLOGY

1048. A POTENTIAL ACCEPTOR OF AMINO GROUPS IN TRANSAMINATING REACTIONS IS:
   A. Glutamine
   B. Glutamic acid
   C. Alanine
   D. Alpha-ketoglutaric acid
   E. Urea
   Ref. p. 817

1049. THE FUNDAMENTAL MOLECULAR PRECURSOR IN THE SYNTHESIS OF UREA IN THE LIVER IS:
   A. Pyruvic acid formed during the glycolytic cycle
   B. Acetyl Coenzyme A
   C. Adenosine-diphosphate (ADP)
   D. Ammonia formed by deamination of amino acids or deamidation of glutamine
   E. Uric acid
   Ref. p. 818

1050. OF ALL OF THE PRECURSORS OF UREA, THE MOST TOXIC IS:
   A. Ornithine
   B. Citrulline
   C. Arginine
   D. Ammonia
   E. Carbamyl phosphate
   Ref. p. 818

1051. THE ORGAN SYSTEM WHOSE FUNCTION IS DISTURBED THE MOST BY ACCUMULATION OF UREA PRECURSORS IS THE:
   A. Liver
   B. Heart
   C. Brain
   D. End plate in neuromuscular junction
   E. Alveolar-capillary membrane
   Ref. p. 818

1052. UREA IS ENZYMATICALLY SPLIT OFF FROM:
   A. Ornithine
   B. Citrulline
   C. Arginine
   D. Carbamyl phosphate
   Ref. p. 818

1053. ALL OF THE FOLLOWING REACTIONS, WITH THE EXCEPTION OF ONE, REQUIRE ENERGY TO PROCEED FORWARD. THIS IS PROVIDED BY THE CONVERSION OF ATP INTO ADP. MARK THE REACTION NOT USING ENERGY DERIVED DIRECTLY FROM ATP:
   A. Hydration of carbon dioxide to form carbonic acid
   B. Efflux of sodium from nerve fibers
   C. Influx of potassium into myocardial fibers
   D. Acto-myosin reaction in sliding myofilaments
   E. Glucose absorption by duodenal mucosa
   Ref. p. 822

1054. HYDROLYSIS OF ATP DRIVES THE FOLLOWING REACTIONS FORWARD (TO THE RIGHT AS WRITTEN BELOW) EXCEPT IN ONE CASE, WHERE ACTUALLY THE REACTION FAVORS THE FORMATION OF NEW ATP (ALTHOUGH IT CAN TAKE PLACE IN BOTH WAYS). MARK THE REACTION WHERE NEW ATP TENDS TO BE FORMED (PREDOMINANT REACTION FROM RIGHT TO LEFT):
   A. Glucose + phosphate ---------- Glucose-6-phosphate
   B. Acetyl Co-A + CO_2 ---------- Malonyl Co-A
   C. Fatty acid + CoA ---------- Fatty acyl-CoA
   D. Adenosine triphosphate ---------- Creatine phosphate
   E. Fructose-6-phosphate ---------- Fructose 1,6-diphosphate
   Ref. p. 822

SECTION VI - METABOLISM, TEMPERATURE REGULATION AND HEPATIC PHYSIOLOGY

1055. THE RELATIVE LACK OF OXYGEN IN TISSUES IS STILL COMPATIBLE WITH A SERIES OF CHEMICAL REACTIONS WHICH, THROUGH THE CATABOLISM OF SUBSTRATES, ALLOWS THE FORMATION OF HIGH ENERGY BONDS. THESE ARE REACTIONS FOUND IN THE UTILIZATION OF:
- A. Carbohydrates
- B. Fats
- C. Proteins
- D. Nucleoproteins
- E. Lipoproteins

Ref. p. 822

1056. THE PREVIOUSLY DESCRIBED ANAEROBIC SEQUENCE OF REACTIONS MAY LAST WITHOUT ANY ADDITION OF OXYGEN MOLECULES FOR A MAXIMUM TIME OF ABOUT:
- A. 2-3 minutes
- B. 5-10 minutes
- C. 10-15 minutes
- D. 25-50 minutes

Ref. p. 823

1057. THE TIME REQUIRED TO "REPAY" AN "OXYGEN DEBT" FOLLOWING PHYSICAL EXERCISE IS:
- A. Related to the time spent doing exercise and not to the amount of work done
- B. That required to resynthesize ATP and creatine phosphate
- C. Required to exhale the excess carbon dioxide accumulated
- D. Determined by the rate of glycolysis and not by the rate of the tricarboxylic acid cycle reactions

Ref. p. 823

1058. IN ENZYMATIC REACTIONS:
- A. The rate is determined solely by enzyme concentrations and not by variations in substrate concentration
- B. The rate is determined solely by variations in substrate concentration and not by enzyme concentration
- C. At maximal substrate concentration the rate is determined only by enzyme concentration
- D. At maximal enzyme concentration the rate does not vary with increasing substrate concentration

Ref. p. 825

1059. FOR A PERSON WHO LIES WITHOUT ANY MOTION EXCEPT FOR BREATHING QUIETLY THE VARIOUS FORMS OF ENERGY ARE CONVERTED INTO:
- A. Pressure
- B. The motion of blood
- C. Digestion and absorption
- D. Mental processes
- E. Heat

Ref. p. 825

1060. MOST IF NOT ALL OF THE POTENTIAL ENERGY CREATED BY THE RHYTHMICALLY CONTRACTING CARDIAC VENTRICLE IS CONVERTED TO HEAT BECAUSE OF:
- A. Molecular friction of blood within vessels
- B. The need to accelerate the blood along vessels
- C. The viscosity of the myocardial fibers as they slide upon each other
- D. The extreme inefficiency of cardiac contractility

Ref. p. 826

SECTION VI - METABOLISM, TEMPERATURE REGULATION AND HEPATIC PHYSIOLOGY

1061. THE BASIC REASON WHY INDIRECT CALORIMETRY IS USED INSTEAD OF THE DIRECT METHOD IS BECAUSE:
   A. There is a very close relationship between oxygen utilized for combustion of substrates and the heat evolved
   B. Direct calorimetry is not as accurate as the indirect method
   C. Heat evolved in the direct method can never be accurately quantitated
   D. Combustion of fat releases a greater amount of heat (calories per liter of oxygen utilized) than combustion of carbohydrates, which cannot be quantitated accurately by the direct method
   Ref. p. 826

1062. ALL OF THE FOLLOWING CALORIC VALUES FOR 1 LITER OF OXYGEN ARE CORRECT EXCEPT ONE. INDICATE WHICH ONE IS THE ERRONEOUS VALUE. THE USE OF 1 LITER OF OXYGEN FOR THE COMBUSTION OF PURE:
   A. Carbohydrates releases 4.0 Kilocalories
   B. Fats releases 4.7 Kilocalories
   C. Proteins releases 4.6 Kilocalories
   D. Ethanol releases 4.8 Kilocalories   Ref. p. 826

1063. IF ONE WERE TO MEASURE OXYGEN CONSUMPTION 2 HOURS AFTER DIFFERENT MEALS IN THE SAME INDIVIDUAL (ALL MEALS HAVING AN ISOCALORIC VALUE), THE HIGHEST INCREASE IN OXYGEN UTILIZATION, DUE TO THE TYPE OF FOOD, WOULD BE ENCOUNTERED IF THE MEAL CONSISTED OF:
   A. Beef steak
   B. Egg yolk
   C. Strawberry cake
   D. Honey
   E. Bread
   Ref. p. 827

1064. A HUMAN MALE WHOSE BASAL METABOLIC RATE PROVES CONSISTENTLY, ON MORE THAN 3 TRIALS, TO BE 5% ABOVE THE MEAN FOR HIS AGE AND BODY BUILD, SHOULD BE CONSIDERED AS BEING:
   A. Hyperthyroid
   B. Hypothyroid
   C. Hypoadrenal
   D. Deficient in hypothalamic function
   E. Probably euthyroid   Ref. p. 828

1065. THE MEASUREMENT OF OXYGEN CONSUMPTION IN AN ANESTHETIZED HUMAN SUBJECT CANNOT BE ACCEPTED AS A MEASURE OF BASAL METABOLIC RATE BECAUSE:
   A. Oxygen consumption is lowered by sleep of any kind, physiological or induced by an anesthetic
   B. Anesthetics tend to alter the rate of oxygen consumption independent of their hypnotic properties
   C. By definition, basal metabolic rate requires that the subject be awake, resting and fasting
   D. It is difficult to compare the oxygen consumption at various levels of anesthesia   Ref. p. 829

SECTION VI - METABOLISM, TEMPERATURE REGULATION   199
               AND HEPATIC PHYSIOLOGY

1066.   THE MOST SERIOUS ALTERATION OF AN ACCURATE BASAL META-
        BOLIC RATE MEASUREMENT IN A PERSON WHO HAS COME TO A
        PHYSICIAN'S OFFICE MAY BE CAUSED BY:
        A. The amount of exercise done in reaching the physician's office
           (climbing stairs, etc.)
        B. The fact that a snack was taken 10 hours before
        C. The slight apprehension caused by the oxygen mask
        D. A room temperature of 80° C   Ref. p. 829

1067.   FROM STUDIES WHICH HAVE DEMONSTRATED THE VARIATION OF
        BASAL METABOLIC RATE (IN KILOCALORIES PER SQUARE METER
        OF SURFACE AREA PER HOUR) AS A FUNCTION OF AGE, ONE MAY
        CONCLUDE THAT:
        A. A unit mass of functional tissues in a 70 year-old male release about
           10 kilocalories less than the same mass in a 10 year-old boy
        B. Older people have a higher rate of tissue oxidations per unit of tis-
           sue mass than a younger person
        C. Older people are basically hypothyroids
        D. The proportion of actively metabolizing tissues in relation to the
           total mass is greater in older people than in younger subjects of the
           same mass and surface area   Ref. Fig. 70-5, p. 828

1068.   THE FOLLOWING DATA WAS OBTAINED IN A LABORATORY WHERE
        THE OXYGEN CONSUMPTION WAS MEASURED IN A MALE SUBJECT.
        Age: 42, Weight:75 Kg; Height: 186 cms; Surface area: 2.0 M^2
        Barometric pressure (corrected) 760 mm Hg; Temperature of gas in
        spirometer = 25° C
        Factor to convert ATPS to STPD = 0.8875   Caloric value of O_2 =
        4.825 Kcal/Liter
        Oxygen consumed/15 minutes = 5 Liters (ATPS)
        Normal BMR at this age (Boothby et al. Am. J. Physiol. 116, 468 (1936)=
        38 Kcal/M^2/hour
        CALCULATIONS REVEAL THAT THIS SUBJECT'S B.M.R. IS:
        A. -20%
        B. -13%
        C. Normal for his age and size
        D. +13%
        E. +20%   Ref. p. 829

        Hint:

$$100\left\{\frac{\left[\frac{(5L \times 4) \times 0.8875 \times 4.825}{2.0}\right] - 38}{38}\right\}$$

SECTION VI - METABOLISM, TEMPERATURE REGULATION AND HEPATIC PHYSIOLOGY

1069. A NUDE PERSON IN A RAFT AT SEA, EXPOSED TO A VERY HOT ENVIRONMENT (BRIGHT HOT SUN AT 36° C, 95% HUMIDITY) WILL PRESERVE HIS CORE TEMPERATURE MOST EFFICIENTLY IF HE:
   A. Stays exposed to the sun without wetting his skin
   B. Shelters himself under some shade and engages in vigorous exercise (rowing)
   C. Wets his skin by periodic immersion in the water
   D. Protects himself from any existing breeze
   Ref. p. 833

1070. IF IN THE PREVIOUS EXAMPLE THE HOT AIR IS BLOWN BY STRONG WINDS AND THE PERSON IS SWEATING, THE LOSS OF HEAT FROM THE HUMAN BODY TO THE ENVIRONMENT TAKES PLACE FOR THE MOST PART BY:
   A. Conduction of heat to the raft
   B. Radiation of heat to the raft and water
   C. Evaporation of sweat secreted over the skin surface
   D. Insensible perspiration   Ref. p. 834

1071. IF IN THE PREVIOUS EXAMPLE THE TEMPERATURE OF THE AIR AND THAT OF THE SEA WATER WERE EXACTLY THE SAME AND THE MAN CHOSE TO SWIM RATHER THAN TO STAY ON THE RAFT:
   A. His heat loss to the water would take place at a faster rate than that of the previous loss to the air
   B. His core temperature would rise very rapidly
   C. His core temperature would not change because the specific heat of water is equal to that of air at this level of temperature
   D. He would not be able to swim because of rapid heat exhaustion
   Ref. p. 833

1072. HUMANS ARE KNOWN TO LOSE ABOUT 0.5 ml WATER/KG BODY WEIGHT/HOUR AS BASAL INSENSIBLE WATER LOSS. KNOWING THE HIGH VALUE OF CALORIC LOSS ENCOUNTERED WHEN 1 ml OF SWEAT WATER (OR PULMONARY ALVEOLAR WATER) IS EVAPORATED AT THE SURFACE OF THE SKIN (OR ALVEOLI), ONE MAY CALCULATE THAT IN 24 HOURS A NON-SWEATING 70 KILOGRAM MAN IN BED IS LOSING:
   A. 287 Kcalories
   B. 387 Kcalories
   C. 487 Kcalories
   D. 587 Kcalories
   E. 687 Kcalories
   Ref. p. 833

   Hint: 0.5 ml/kg/hr x 70 kg x 24 hrs x 0.58 Kcal/ml

1073. A REDUCTION IN PLASMA VOLUME CAUSED BY EXCESSIVE SWEATING IS FOLLOWED BY INCREASED ALDOSTERONE SECRETION BY ADRENAL GLANDS. THIS WILL:
   A. Increase the core temperature because of an effect of steroids in general on hypothalamic thermal-sensitive neurons
   B. Decrease the core temperature because of vasoconstriction of skin arterioles secondary to an impending circulatory failure
   C. Increase the volume flow of water in sweat
   D. Reduce the concentration of sodium and raise the concentration of potassium in sweat   Ref. p. 835

SECTION VI - METABOLISM, TEMPERATURE REGULATION   201
              AND HEPATIC PHYSIOLOGY

1074. TO PRESERVE A NORMAL CORE TEMPERATURE IN THE HOT
       DESERT, THE BEST CLOTHING IS THAT WHICH:
       A. Decreases reflection of radiant heat from clothing to the environment
       B. Increases reflection of radiant heat back to the environment (white
          clothing)
       C. Favors sweating but preserves a layer of moist air close to the body
          (nylon)
       D. Prevents evaporation of sweating   Ref. p. 837

1075. TEMPERATURE ELEVATIONS BEYOND 37º C IN MINUTE AREAS OF
       THE HYPOTHALAMUS ARE FOLLOWED BY RESPONSES THAT TAKE
       PLACE IN THE WHOLE BODY AND WHICH FAVOR THE OVERALL
       LOSS OF HEAT. THESE AREAS ARE LOCATED IN THE:
       A. Mamillary bodies   D. Dorsomedial nucleus
       B. Median eminence   E. Preoptic area
       C. Ventromedial nucleus   Ref. p. 838

1076. IN A GRAM NEGATIVE BACTERIAL INFECTION AN ENDOTOXIN
       THAT HAS BEEN JUST RELEASED INTO THE BLOOD STREAM WILL
       CAUSE THE HYPOTHALAMIC CENTERS CONCERNED WITH THE
       REGULATION OF TEMPERATURE TO:
       A. Reset to a lower temperature threshold
       B. Reset to a higher temperature threshold
       C. Cause in some manner the subjective sensation that the body is too
          hot
       D. Decrease the heat production mechanisms
                                             Ref. p. 841

1077. A USEFUL PHYSIOLOGICAL COMPENSATION FOUND IN A PERSON
       WHO HAS BECOME ACCLIMATIZED TO HEAT IS THAT:
       A. He learns to avoid the causes that increase heat production
       B. There is an increase in the percentage of heat lost through radiation
       C. He develops a new hypothalamic threshold to heat
       D. The adrenal cortices secrete more aldosterone
                                             Ref. p. 842

1078. NOT TO BE CONFUSED WITH THE CALORIC VALUE OF 1 LITER OF
       OXYGEN USED FOR THE COMBUSTION OF EACH OF THE THREE
       MAIN TYPES OF FOOD (CARBOHYDRATE, LIPID, AND PROTEIN) IS
       THE CALORIC VALUE OF THE UNIT MASS (1 GRAM) OF SUCH FOODS.
       THUS, THE TOTAL COMBUSTION (IN A BOMB CALORIMETER) OF
       1 GRAM OF PURE FAT YIELDS ABOUT:
       A. 1.3 Kilocalories/gram   D. 15.3 Kilocalories/gram
       B. 4.3 Kilocalories/gram   E. 21.3 Kilocalories/gram
       C. 9.3 Kilocalories/gram   Ref. p. 844

1079. FATTY FOODS HAVE A HIGH PERCENTAGE OF FAT AND RELATIVE-
       LY LITTLE OF WATER OR ANYTHING ELSE. THUS, BUTTER (OF
       ANIMAL ORIGIN) CONTAINS APPROXIMATELY:
       A. 11 grams of fat/100 grams of butter
       B. 31 grams of fat/100 grams of butter
       C. 61 grams of fat/100 grams of butter
       D. 81 grams of fat/100 grams of butter
       E. 91 grams of fat/100 grams of butter Ref. Table 72-1, p. 845

SECTION VI - METABOLISM, TEMPERATURE REGULATION AND HEPATIC PHYSIOLOGY

1080. WHEN WHOLE MILK IS "SKIMMED" AND ITS FAT REMOVED, THE FAT CONTENT DECREASES FROM:
A. 10% before skimming to less than 2% in skimmed milk
B. 7.9% before skimming to less than 1% in skimmed milk
C. 5.9% before skimming to less than 1% in skimmed milk
D. 3.9% before skimming to less than 1% in skimmed milk
E. 1.9% before skimming to less than 1% in skimmed milk
Ref. p. 845

1081. AN ADULT HUMAN WHO EXCRETES 10 GRAMS OF NITROGEN PER DAY IN THE URINE HAS A TOTAL URINARY PROTEIN EXCRETION EQUAL TO ABOUT:
A. 12 grams per day
B. 32 grams per day
C. 62 grams per day
D. 102 grams per day
E. 252 grams per day
Ref. p. 846

Hint: $10 \times \frac{100}{16}$

1082. IN ORDER TO BALANCE THE PROTEIN LOSSES DESCRIBED IN THE PREVIOUS QUESTION, THE SAME SUBJECT MUST EAT EVERY DAY A MINIMUM OF:*
A. 264 grams of meat
B. 364 grams of meat
C. 464 grams of meat
D. 564 grams of meat
E. 664 grams of meat
Ref. Table 72-1, p. 845

*Assume the meat contains 17 grams of protein per cent.

1083. A DIABETIC WHO BECOMES DECOMPENSATED AND LAPSES INTO COMA SHOULD SHOW APPROXIMATELY THE FOLLOWING RATIO OF $\frac{\text{EXPIRED } CO_2/\text{MINUTE}}{O_2 \text{ CONSUMED}/\text{MINUTE}}$.

A. $\frac{500 \text{ ml } CO_2/\text{min}}{400 \text{ ml } O_2/\text{min}}$

B. $\frac{600 \text{ ml } CO_2/\text{min}}{500 \text{ ml } O_2/\text{min}}$

C. $\frac{400 \text{ ml } CO_2/\text{min}}{500 \text{ ml } O_2/\text{min}}$

D. $\frac{580 \text{ ml } CO_2/\text{min}}{800 \text{ ml } O_2/\text{min}}$
Ref. p. 846

1084. THE INCREASED UTILIZATION OF CARBOHYDRATES TO SYNTHESIZE NEW TRIGLYCERIDES RESULTS IN A RESPIRATORY QUOTIENT OF ABOUT:
A. 1.2
B. 1.0
C. 0.82
D. 0.71
Ref. p. 847

SECTION VI - METABOLISM, TEMPERATURE REGULATION   203
                AND HEPATIC PHYSIOLOGY

1085.   IT IS BELIEVED THAT AN ANIMAL MAY CEASE TO EAT BECAUSE
        OF A TOTAL LACK OF THE NORMAL HUNGER DRIVE IF THE CEN-
        TRAL NERVOUS SYSTEM IS DESTROYED IN A VERY SMALL LOCAL-
        IZED AREA PLACED IN THE:
        A. Preoptic region of the hypothalamus
        B. Ventromedial nucleus of the hypothalamus
        C. Mamillary bodies
        D. Hypothalamic-pituitary portal vessels
        E. Lateral hypothalamic area   Ref. p. 847

1086.   A MESENCEPHALIC ANIMAL, I.E. ONE WHOSE BRAIN STEM HAD
        BEEN TRANSECTED AT A HIGH LEVEL IN THE CEREBRAL
        PEDUNCLES, BUT WHOSE NEURAL CONNECTIONS BETWEEN
        HYPOTHALAMUS AND MORE CAUDAL SEGMENTS HAD BEEN IN-
        TERRUPTED WILL:
        A. Be capable of chewing food
        B. Still demonstrate behavioral responses to appetite
        C. Be totally incapable of swallowing
        D. Have lost all salivary secretions   Ref. p. 847

1087.   BESIDES THE HYPOTHALAMIC CENTERS BELIEVED TO CONTROL
        APPETITE OR HUNGER:
        A. The rest of the cerebrum has no role to play in this control
        B. The premotor area of the frontal cortex assists in controlling the
           intensity of the hunger drive
        C. There is an additional role played by the limbic cortex
        D. The cerebellar flocculus participates in some of the hunger
           responses   Ref. p. 848

1088.   ALL OF THE FOLLOWING BLOOD CHANGES IN EXPERIMENTAL
        ANIMALS HAVE BEEN SHOWN TO BE ASSOCIATED WITH A DIMINU-
        TION IN THE SPONTANEOUS DRIVE TO SEARCH FOR FOOD, EXCEPT:
        A. Hyperglycemia
        B. Increase in amino acid concentration in the blood
        C. Increase in blood temperature
        D. Gastric distention
        E. Hypoglycemia   Ref. p. 848

1089.   DURING TOTAL STARVATION, PHYSIOPATHOLOGICAL CHANGES DUE
        TO RELATIVE AVITAMINOSIS WILL APPEAR IN SEQUENCE AS BODY
        STORES ARE EXHAUSTED. THE EARLIEST AVITAMINOSIS WILL BE
        THAT DUE TO THE RELATIVE LACK OF:
        A. Vitamin A
        B. Vitamin B_1
        C. Vitamin D
        D. Vitamin C   Ref. p. 852

1090.   AN ABNORMAL KERATINIZATION OF THE CORNEA CAN OCCUR AS
        A CONSEQUENCE OF THE RELATIVE DEFICIT OF:
        A. Vitamin A
        B. Vitamin B_1 (thiamine)
        C. Nicotinic acid
        D. Vitamin B_6 (pyridoxine)
        E. Vitamin E   Ref. p. 853

SECTION VI - METABOLISM, TEMPERATURE REGULATION AND HEPATIC PHYSIOLOGY

1091. A CLINICAL HISTORY OF NUMBNESS AND TINGLING IN THE EXTREMITIES PLUS ARREFLEXIA AND MUSCULAR WEAKNESS IN A PERSON WHO HAS BEEN UNDERNOURISHED SHOULD SUGGEST THE DIAGNOSIS OF RELATIVE DEFICIENCY OF:
    A. Vitamin A
    B. Vitamin B_1
    C. Vitamin D
    D. Vitamin C   Ref. p. 853

1092. A PROMINENT FEATURE OF THIAMINE DEFICIENCY IS A(N):
    A. Accelerated venous return to the right atrium and right ventricle
    B. Low cardiac output
    C. Low end diastolic right ventricular pressure
    D. High value of the hematocrit   Ref. p. 853

1093. A DIET RICH IN CORN BUT POOR IN MANY OTHER RESPECTS, INCLUDING THE VARIOUS B VITAMINS, IS LIKELY TO LEAD TO PELLAGRA BECAUSE IT:
    A. Is very difficult to digest
    B. Lacks methionine
    C. Is poor in tryptophan, a precursor of nicotinic acid
    D. Leads to steatorrhea   Ref. p. 854

1094. RIBOFLAVIN (OR VITAMIN B_2) IS ESSENTIAL TO BODY METABOLISM BECAUSE IT IS A MOLECULAR COMPONENT OF:
    A. DPN   D. FAD
    B. TPN   E. Vitamin B_{12}
    C. Coenzyme A   Ref. p. 854

1095. THE FORMATION OF PURINES AND THYMINE IS ESSENTIALLY LINKED TO PROTEIN SYNTHESIS AND SEEMS TO BE DEPENDENT, AMONG OTHER THINGS, ON THE NORMAL AVAILABILITY OF:
    A. Vitamin A
    B. Thiamine
    C. Pantothenic acid
    D. Pteroylglutamic acid (folic acid)
    E. Biotin   Ref. p. 855

1096. A LOW CONVULSIVE THRESHOLD IN HUMANS MAY BE AN INDICATION OF DEFICIENCY OF:
    A. Vitamin A   D. Pantothenic acid
    B. Vitamin B   E. Folic acid
    C. Pyridoxine   Ref. p. 855

1097. WHILE EXPERIMENTAL ANIMALS MAY BECOME DEFICIENT, BY EXPERIMENTAL DESIGN, IN EVERY VITAMIN KNOWN, THERE IS NO RECORD THAT HUMANS MAY BECOME DEFICIENT IN:
    A. Folic acid
    B. Pantothenic acid
    C. Biotin
    D. Nicotinic acid
    E. Vitamin B_{12} Ref. p. 856

SECTION VI - METABOLISM, TEMPERATURE REGULATION   205
                 AND HEPATIC PHYSIOLOGY

1098.   A CRITICAL SHORTAGE OF INTERCELLULAR SUBSTANCE IN BONE
        IS A CHARACTERISTIC FEATURE OF THE DEFICIENCY IN:
        A. Vitamin A   D. Vitamin D
        B. Vitamin B_1 E. Vitamin E
        C. Vitamin C   Ref. p. 857

1099.   VITAMIN K:
        A. Is stored in relatively large quantities in the liver
        B. Must be ingested daily or at frequent intervals, as the body is incapable of synthesizing its structure
        C. Deficiency occurs readily if colonic bacteria are eliminated by the administration of an antibiotic
        D. Is required in the sequence of events that precedes the formation of thromboplastin   Ref. p. 858

1100.   AMONG THE FOLLOWING MINERAL COMPONENTS OF THE HUMAN
        BODY, THE LEAST ABUNDANT IS:
        A. Iodine   D. Iron
        B. Sulfur   E. Calcium
        C. Magnesium   Ref. Table 73-2, p. 858

1101.   CALCIUM AND MAGNESIUM IONS HAVE SOME COMMON PHYSIO-
        LOGICAL PROPERTIES, WHILE ON THE OTHER HAND THEY SEEM
        TO ANTAGONIZE EACH OTHER IN OTHER SITUATIONS. THUS:
        A. Somnolence and torpor may be caused by an excess of either calcium or magnesium
        B. The deficit of calcium causes tetany while the deficit of magnesium causes muscle flaccidity but never tetany
        C. The deficit of calcium causes excitability of the central nervous system while the deficit of magnesium has no effect on the central nervous system
        D. The effects of an excess of magnesium on skeletal muscle cannot be reversed by an injection of calcium
                                  Ref. p. 859

1102.   CYTOCHROME C HAS A PORPHYRIN RING BOUND TO TWO IMIDAZOLE
        GROUPS AND TO A PROTEIN. THE ELEMENT WHICH IS CO-
        ORDINATED TO THE PORPHYRIN RING IS:
        A. Magnesium   D. Oxygen
        B. Sulfur   E. Iron
        C. Calcium   Ref. p. 859

1103.   ZINC IS AN ESSENTIAL COMPONENT OF ALL THE FOLLOWING
        ENZYMES, EXCEPT:
        A. Glutamic aspartic transaminase
        B. Carbonic anhydrase in red blood cells
        C. Lactic acid dehydrogenase
        D. Carboxypeptidase
        E. Carbonic anhydrase in renal tubular cells
                                  Ref. p. 859

SECTION VI - METABOLISM, TEMPERATURE REGULATION AND HEPATIC PHYSIOLOGY

1104. THE ADMINISTRATION OF COBALT TO EXPERIMENTAL ANIMALS IN SMALL DOSES HAS RESULTED IN THE DEVELOPMENT OF:
A. High values of mean hemoglobin concentration in red cells
B. Sluggish erythropoiesis
C. Slow development of platelet precursors
D. Polycythemia
E. Anemia   Ref. p. 859

1105. A SUCCESSFUL PROGRAM OF ADMINISTRATION OF FLUORIDE DESIGNED TO PREVENT CARIES OF THE TEETH SHOULD STRESS THE TIMING OF THIS PROPHYLACTIC THERAPY, ESPECIALLY:
A. During the prenatal care, giving fluoride to the mother in the first months of pregnancy
B. During the immediate postnatal period and during the time when teeth are being formed
C. Between 10-15 years of life
D. Between 15-20 years of life   Ref. p. 860

1106. PRESSURE WITHIN THE CENTRAL VEIN AND SINUSOIDS OF THE LIVER PARENCHYMA IS LIKELY TO RISE SIGNIFICANTLY:
A. If the extrahepatic portion of the portal vein is compressed by lymph nodes
B. If diagnostic doses of bromsulphalein are injected
C. After fatty meals
D. If contractility of the right ventricle is reduced
E. If a person assumes a standing position after lying supine in bed
                                                        Ref. p. 862

1107. FREE FLUID ACCUMULATED IN THE PERITONEAL CAVITY:
A. Is more likely to occur with increased pressure in the intrahepatic than in extrahepatic capillaries, such as those present in the intestinal wall
B. Never has a protein concentration higher than in plasma
C. Is always associated with a slow rate of hepatic lymph flow
D. Does not exceed a value of about 1.5-2 liters
                                                        Ref. p. 862

1108. ALTHOUGH THE VOLUME OF THE DISTENDED HUMAN ADULT GALLBLADDER DOES NOT EXCEED ABOUT 50 ml, A GREATER VOLUME OF DILUTE BILE IS CONTINUOUSLY SECRETED BY THE LIVER. IN 24 HOURS, THE TOTAL VOLUME OF BILE SECRETED BY THE LIVER AMOUNTS TO ABOUT:
A. 100-200 ml
B. 500-600 ml
C. 1000-2000 ml
D. 5000-6000 ml   Ref. p. 863

SECTION VI - METABOLISM, TEMPERATURE REGULATION AND HEPATIC PHYSIOLOGY

1109. THE THREE-FOLD CONCENTRATION OF BILE SALTS AND CHO-
LESTEROL THAT TAKES PLACE IN GALLBLADDER BILE, WHILE IT
ACCUMULATES IN THIS VISCUS BETWEEN MEALS, WOULD BE
OSMOTICALLY INTOLERABLE IF OTHER COMPONENTS WERE NOT
REABSORBED ALONG WITH THE BULK OF WATER. THUS, THE
SODIUM CONCENTRATION DECREASES FROM ABOUT:
   A. 145 mEq/L in hepatic bile to about 130 mEq/L in gallbladder bile
   B. 145 mEq/L in hepatic bile to about 50 mEq/L in gallbladder bile
   C. 145 mEq/L in hepatic bile to about 10 mEq/L in gallbladder bile
   D. 145 mEq/L in hepatic bile to no sodium in gallbladder bile
   Ref. Table 74-1, p. 863

1110. A KEY DETERMINANT OF THE VOLUME OF BILE SECRETED BY
THE LIVER EACH DAY IS THE:
   A. Rate of conjugation of bilirubin
   B. Number of meals ingested per day
   C. Volume of the gallbladder when it is fully distended
   D. Amount of bile salts recycled through the enterohepatic circulation
   E. Concentration of cholesterol in bile
   Ref. p. 864

1111. THE FUNDAMENTAL TRANSFORMATION OF BILIRUBIN IN THE
LIVER CONSISTS OF:
   A. Its attachment to an albumin molecule
   B. The opening of the heme ring
   C. The conjugation with glucuronic acid
   D. The passive diffusion through the large endothelial pores and excretion of a basically intact molecule as presented by the incoming portal blood
   Ref. p. 864

1112. UROBILINOGEN FORMED IN THE INTESTINAL LUMEN FROM
BILIRUBIN:
   A. Is reabsorbed in toto into the portal blood
   B. Cannot pass the renal glomerular capillary membranes
   C. Must be conjugated in order to circulate in blood
   D. Is not found in urine if the common bile duct is totally occluded
   Ref. p. 866

1113. THE MAJOR URINARY FORM OF BILIRUBIN OR ITS METABOLITES IN
SEVERE OBSTRUCTIVE JAUNDICE IS:
   A. Unconjugated bilirubin
   B. Urobilinogen
   C. Bilirubin conjugated with glucuronic acid
   D. Stercobilin
   Ref. p. 866

1114. IN THE SO-CALLED VAN DEN BERGH TEST IN BLOOD:
   A. The "direct" reaction occurs rapidly after adding reagents, because most of the blood's bilirubin is conjugated
   B. A high concentration and preponderance of the "indirect" type over the "direct" type is compatible with obstruction to bile flow in intrahepatic canaliculi
   C. A high concentration and preponderance of the "direct" type is diagnostic of an increased rate of destruction of erythrocytes (hemolysis)
   D. The albumin-bound bilirubin is the most soluble form and reacts earlier with the reagents
   Ref. p. 866

SECTION VI - METABOLISM, TEMPERATURE REGULATION AND HEPATIC PHYSIOLOGY

1115. OF THE FOLLOWING VITAMINS THE ONE WHICH IS STORED IN GREATEST ABUNDANCE IN THE LIVER IS:
A. Vitamin A
B. Vitamin B_1
C. Vitamin B_2
D. Vitamin D

Ref. p. 868

SECTION VII - ENDOCRINE PHYSIOLOGY

IN THE FOLLOWING QUESTIONS (1116-1225) A STATEMENT IS FOLLOWED BY FOUR POSSIBLE ANSWERS. ANSWER BY USING THE KEY OUTLINED BELOW:
A. If only 1, 2 and 3 are correct
B. If only 1 and 3 are correct
C. If only 2 and 4 are correct
D. If only 4 is correct
E. If all are correct

1116. 3'5'CYCLIC ADENOSINE-MONOPHOSPHATE (CYCLIC AMP):
1. Is formed by an enzyme whose substrate is ATP
2. Is basically formed at the membrane of cells or as a consequence of a membrane attached enzyme
3. Mediates the action of ACTH in adrenal cortical cells
4. Is known to mediate the action of growth hormone
   Ref. p. 873

1117. SUCCESSFUL TRANSPLANTATION OF THE WHOLE PITUITARY TO THE RENAL CAPSULE OF AN EXPERIMENTAL ANIMAL:
1. Will result in secretion of vasopressin by the transplanted posterior pituitary in some well operated cases
2. Will result in a limited but variable secretion of TSH (thyrotropin)
3. Will never be followed by recanalization of hypothalamic-pituitary portal veins if the pituitary is transplanted back under the median eminence
4. Abolishes completely the hypothalamic influence over the anterior pituitary if it remains within the area of the renal capsule
   Ref. p. 875

1118. ALTHOUGH THERE IS NOT ABSOLUTE CERTAINTY ABOUT THE EXACT LOCATION WITHIN THE HYPOTHALAMUS WHERE EACH OF THE VARIOUS HYPOTHALAMIC RELEASING FACTORS ARE SYNTHESIZED, MUCH EVIDENCE SUGGESTS THAT THE <u>PREOPTIC</u> AREA IS PROBABLY CONCERNED IN PART WITH:
1. Secretion of thyrotropin-releasing-factor
2. Secretion of growth hormone-releasing-factor
3. Part of the controlling mechanism which influences the secretion of FSH and LH releasing factors
4. Secretion of ADH   Ref. p. 876

1119. A YOUNG BOY WHO RECEIVES SLIGHTLY EXCESSIVE THERAPEUTIC DOSES OF GROWTH HORMONE MAY EXHIBIT:
1. Hyperglycemia
2. An accelerated turnover of fatty acids
3. A positive nitrogen balance
4. A wider epiphyseal cartilage plate   Ref. p. 877

SECTION VII - ENDOCRINE PHYSIOLOGY

1120. IN RELATION TO EFFECTS OF GROWTH HORMONE ON PROTEIN SYNTHESIS, IT HAS BEEN SHOWN EXPERIMENTALLY THAT, AFTER GROWTH HORMONE INJECTIONS:
1. Transport of numerous amino acids through cell membranes is accelerated above values found in controls or hypophysectomized animals
2. Messenger RNA is increased
3. Cell free systems can still show positive effects of growth hormone provided the injections of the hormone took place in the living animal
4. There is no effect caused by growth hormone on the role of ribosomes
Ref. p. 878

1121. THE HYPERGLYCEMIC EFFECT OF GROWTH HORMONE INJECTIONS:
1. Is due in part to inhibition of insulin-mediated facilitated diffusion of glucose into adipose tissue cells and skeletal muscle fibers
2. May be overcome by small doses of insulin
3. May be increased if cortisol is given at the same time in high doses
4. May be overcome by ACTH   Ref. p. 878

1122. A HIGH CONCENTRATION OF GROWTH HORMONE IN BLOOD MAY BE SEEN:
1. During starvation
2. In acromegaly
3. In gigantism
4. Following insulin induced hypoglycemia
Ref. p. 879

1123. AN 8 YEAR-OLD PATIENT WHOSE HYPOPHYSIS HAS BEEN DESTROYED IN PART OR IN TOTO BY A COMPRESSING TUMOR WITHIN THE SELLA TURCICA:
1. Will have a normal growth potential if he is treated with exogenous growth hormone of bovine origin
2. May develop easily a hyperglycemia after meals rich in carbohydrates because of the lack of growth hormone
3. Should not receive cortisol as treatment under any circumstance
4. Is likely to be deficient in thyroid and cortisol secretions
Ref. p. 880

1124. A 24 YEAR-OLD WOMAN WHO DEVELOPS AN ACIDOPHILIC ADENOMA OF THE ANTERIOR PITUITARY:
1. Will have continued growth of metaphysial cartilage due to the excess of growth hormone
2. Is in greater danger of becoming a diabetic than a person who does not have such a lesion
3. Does not usually respond to radiotherapy of the pituitary tumor
4. Will likely have to buy a larger pair of shoes than those used before the disease started   Ref. p. 880

1125. INCLUDED AMONG BONES WHICH UNDERGO STRIKING ENLARGEMENT IN CASES OF ACROMEGALY ARE THE:
1. Femur in the longitudinal axis
2. Supraorbital ridges (frontal bone)
3. Humerus in the longitudinal axis
4. Vertebral bodies   Ref. p. 880

SECTION VII - ENDOCRINE PHYSIOLOGY

1126. THE TRANSECTION OF THE LOWER PORTION OF THE PITUITARY STALK:
1. Is not likely to cause diabetes insipidus if there is no injury to the median eminence
2. Will produce hypogonadism if the portal vessels are not allowed to regenerate
3. Will be followed by increased sensitivity to insulin in a previously treated diabetic
4. Will require that thyroid hormone be given therapeutically if a low protein bound iodine develops a few months after the transection
Ref. p. 882

1127. ANTIDIURETIC HORMONE (VASOPRESSIN), SO VITAL IN THE NORMAL REGULATION OF BODY WATER AND OSMOLARITY, IS KNOWN TO:
1. Act on the distal tubular cells of the nephron through the increased formation of cyclic-AMP in these cells
2. Have a half-life of about 24 hours
3. Originate as a product of neuronal metabolism in the supraoptic nuclei of the hypothalamus
4. Be secreted at a continuous rate throughout the day and night
Ref. p. 883

1128. THE ACTIVE RELEASE OF OXYTOCIN:
1. Takes place from the anterior pituitary when there is a (reflex) tactile stimulation of the female lactating breast
2. Causes the pregnant uterus to contract
3. Originates from cells located in the ventromedial nucleus of the hypothalamus
4. Is inhibited by a hypothalamic factor in the non-lactating female
Ref. p. 884

1129. IN RELATION TO THE PHYSIOLOGICAL ROLE OF THE PINEAL GLAND, IT IS APPARENTLY TRUE THAT:
1. The pineal gland in adult humans is calcified in most cases
2. The calcification of the pineal gland, which is equivalent to the gland's removal, has no known effect on the body as a whole
3. "Melatonin" extracted from pineals causes the skin of the frog to become lighter
4. "Melatonin" is believed to inhibit the secretion of pituitary gonadotropins in rats
Ref. p. 884

1130. THE ADMINISTRATION OF HIGH DOSES OF ALDOSTERONE TO A MAMMAL (RAT, DOG, HUMAN):
1. Will cause a significant increase in the sodium reabsorption by the distal tubular cells of the renal nephrons
2. Will cause little if any changes in the sodium transport by the proximal tubular cells of the renal nephrons
3. May increase the reabsorption of not more than 10-15% of the filtered sodium
4. May increase the reabsorption of more than 50% of the filtered sodium
Ref. p. 887

SECTION VII - ENDOCRINE PHYSIOLOGY

1131. CONCOMITANT WITH EVENTS LEADING TO SODIUM REABSORPTION IN DISTAL TUBULES OF RENAL NEPHRONS THERE IS:
    1. Increased reabsorption of hydrogen ions in the distal tubules
    2. Increased tubular secretion of potassium ions by the distal tubular cells
    3. A slight shift of the plasma in the peritubular capillaries to a more acid pH than that found in arterial plasma
    4. An increased return of bicarbonate ions to the extracellular fluid
    Ref. p. 888

1132. AS A CONSEQUENCE OF AN EXCESSIVE SECRETION OF ALDOSTERONE FROM A FUNCTIONAL ADRENAL CORTICAL TUMOR, A PATIENT MAY SHOW:
    1. Hypertension
    2. Alkalosis
    3. Hypokalemia
    4. An increased inulin space   Ref. p. 888

1133. THE MECHANISM THROUGH WHICH ALDOSTERONE CAUSES THE REABSORPTION OF SODIUM BY THE RENAL TUBULE:
    1. Obtains its effects within less than 2 minutes after the hormone has reached the kidney
    2. Involves a sequence of chemical transformations limited to the membrane of tubular cells
    3. Is basically one of facilitated diffusion
    4. Is linked to protein synthesis in the renal tubular cells
    Ref. p. 889

1134. IT HAS BEEN ESTABLISHED THAT ALDOSTERONE SECRETION IS INDUCED TO PROCEED AT A FASTER RATE WHEN:
    1. The concentration of sodium in the plasma that perfuses the adrenal cortex has decreased below a certain value
    2. The concentration of potassium in the plasma that perfuses the adrenal cortex has decreased below a certain value
    3. The juxtaglomerular cells form more renin, leading to a higher angiotensin titer in blood
    4. Corticotrophin (ACTH) secretion from the anterior pituitary decreases
    Ref. p. 890

1135. INCREASED CONCENTRATIONS OF CORTISOL IN BLOOD CAUSE INCREASED:
    1. Catabolism of proteins in bones and muscle
    2. Concentration of amino acids in blood
    3. Protein synthesis in the liver
    4. Deamination of amino acids in liver cells
    Ref. p. 892

1136. A PATIENT WITH A FUNCTIONAL TUMOR OF THE ADRENAL CORTEX WHICH PRODUCES AN EXCESS OF CORTISOL SHOWS:
    1. Increased catabolism of triglycerides in adipose tissue
    2. Increased concentrations of free fatty acids in plasma
    3. Increased deposition of fat in the trunk
    4. A high rate of ACTH secretion   Ref. p. 892

SECTION VII - ENDOCRINE PHYSIOLOGY

1137. ALTHOUGH THE EXACT SIGNIFICANCE OF MANY OF THE ANTI-INFLAMMATORY EFFECTS OF CORTISOL IS UNKNOWN, IT IS TRUE THAT AN EXCESS OF THIS STEROID IN BLOOD WILL:
   1. Inhibit formation of antibodies originating in lymphocytes
   2. Prevent the antigen-antibody reaction
   3. Prevent release of proteolytic enzymes from lysosomes
   4. Accelerate diapedesis of leukocytes
   Ref. p. 894

1138. A PATIENT WHO HAS HAD DESTRUCTION OF HIS ADRENAL GLANDS PRODUCED BY BILATERAL METASTASIS FROM A BRONCHOGENIC CARCINOMA, WILL, IF HE IS NOT TREATED:
   1. Have a high hematocrit
   2. Develop a low total body sodium
   3. Have elevated plasma concentrations of ACTH
   4. Have a low glomerular filtration rate
   Ref. p. 898

1139. HYPERGLYCEMIA IN THE EARLY STAGES OF A CASE OF CUSHING'S DISEASE IS MAINLY DUE TO:
   1. Decreased insulin secretion
   2. Increased secretion of growth hormone
   3. Decreased glomerular filtration rate
   4. Increased gluconeogenesis in the liver
   Ref. p. 891

1140. THE DIAGNOSIS OF A TUMOR OF THE ADRENAL CORTEX WHICH SECRETES ANDROSTENEDIONE AND DEHYDROEPIANDROSTERONE AT HIGH RATES:
   1. Is easier to make in an adult female than in an adult male
   2. Will cause gynecomastia
   3. Will cause the excretion of large quantities of 17-ketosteroids in the urine
   4. Is literally impossible to make in a young child (6 months of age)
   Ref. p. 900

1141. THE TOTAL ABLATION OF THE THYROID GLAND, NOT FOLLOWED BY ANY REPLACEMENT THERAPY, WILL CAUSE IN AN ADULT HUMAN:
   1. A fall of the basal metabolic rate to levels equal to -90%
   2. A low cardiac output
   3. Low thyrotropin levels in blood
   4. A poor tolerance to cold weather   Ref. pp. 902-913

1142. TYROSINE, THE AMINO ACID WHICH SERVES AS THE BASIS FOR THE SYNTHESIS OF THYROXINE:
   1. Cannot be iodinated by iodide (in ionic form) but only by iodine
   2. Exists in the thyroid cell as an integral part of a large protein
   3. May be synthesized in the human body
   4. Is largely released to the circulation as a diiodotyrosine
   Ref. p. 903

SECTION VII - ENDOCRINE PHYSIOLOGY

1143. THYROTROPIN, A PITUITARY HORMONE THAT REGULATES THE RATE OF TURNOVER OF THYROID HORMONES:
   1. May increase the "iodide trap" five-fold
   2. Increases the activity of the proteinases which split thyroxine from thyroglobulin
   3. Is secreted at a much slower rate than normally if the concentration of thyroxine blood levels increases
   4. Is secreted at a higher rate than normal when the body temperature tends to decrease, because of a moderately cold environment
   Ref. pp. 907, 908

1144. THYROTROPIN:
   1. Does not increase the levels of cyclic AMP in thyroid cells
   2. Is positively known to be responsible for the exophthalmos seen in cases of hyperthyroidism
   3. Is secreted at much higher rates when there is an acute severe stress
   4. Is secreted at a much higher rate when synthesis of thyroxine is inhibited, as in patients receiving propylthiouracil
   Ref. p. 908

1145. PROBABLE REASONS WHY TRIIODOTHYRONINE ACTS IN SMALLER DOSES AND EARLIER THAN THYROXINE ARE:
   1. Triiodothyronine has little affinity for thyroxine-binding globulin (TBG)
   2. Thyroxine has a greater affinity for thyroxine-binding globulin
   3. The bound hormone is in equilibrium with the free fraction; the free hormone fraction is in lower concentration in the case of thyroxine
   4. The free fractions are believed to be the active form of the hormones
   Ref. p. 904

1146. WHEN ONE COMPARES THE OXYGEN CONSUMPTION (IN VITRO) OF SEVERAL TISSUES IN THE BODY OF AN ANIMAL WHICH HAS RECEIVED THYROID HORMONE, FOR SEVERAL DAYS PRIOR TO SACRIFICE, ONE MAY FIND THAT THE:
   1. Heart has a greater increment in the consumption of oxygen than that shown by the spleen
   2. Brain shows little or no change compared to controls
   3. Retina has little or no change in O_2 consumption when compared to controls
   4. Testes show a remarkable increase in oxygen consumption
   Ref. p. 904

1147. THE ROLE PLAYED BY THYROID HORMONE ON FAT METABOLISM IS OF SUCH A NATURE THAT A SLIGHT EXCESS CAUSES A(N):
   1. Increased turnover of body fats
   2. Decrease in blood cholesterol
   3. Increased deposition of fat if an excess of carbohydrates is ingested
   4. Diminished lipase activity in fat depots
   Ref. p. 905

1148. AN EXCESSIVE SECRETION OF THYROXINE IS LIKELY TO CAUSE:
   1. Constipation, due to low motility of the gastrointestinal tract
   2. A great increase in the rate of conduction of nerve impulses
   3. Somnolence
   4. Deficiency of thiamine
   Ref. pp. 906, 907

SECTION VII - ENDOCRINE PHYSIOLOGY

1149. THE HYPOTHALAMIC REGULATION OF THE RELEASE OF THYROTROPIN IS OF SUCH A NATURE THAT:
1. Without the normal function of hypothalamic releasing factors (removed by electrolytic lesion of small nuclei in the hypothalamus) the release of TSH drops to zero
2. The pituitary retains a great deal of autonomy if it is separated from the hypothalamus
3. Thyrotropin secreting cells in the pituitary no longer respond to fluctuations in thyroxine blood levels if the pituitary is transplanted to other areas of the body
4. Purely emotional situations do influence the rate of secretion of TSH
Ref. p. 908

1150. DECREASED TRANSPORT OF IODIDE FROM PLASMA INTO THYROID CELLS MAY BE CAUSED BY:
1. The administration of thiocyanates
2. The administration of perchlorates
3. The excess of iodide in plasma
4. Propylthiouracil
Ref. p. 909

1151. THERAPEUTIC ADMINISTRATION OF PROPYLTHIOURACIL CAUSES:
1. An inhibition of synthesis of thyroxine
2. An inhibition of the trapping mechanism of iodide anions from blood
3. Goiter
4. A decreased perfusion of the thyroid gland
Ref. p. 909

1152. SUCCESSFUL SURGICAL THERAPY OF HYPERTHYROIDISM REQUIRES THAT:
1. Euthyroidism be approached by inhibiting the excessive rate of synthesis of thyroid hormone
2. The patient not receive any iodide prior to surgery
3. Lugol solution or a similar iodide preparation be added to the treatment in the last few weeks prior to surgery
4. The basal metabolic rate does not fall below +35%
Ref. p. 911

1153. IT IS TYPICAL OF A PATIENT WHO IS HYPERTHYROID TO:
1. Have invariably an exophthalmos
2. Have a goiter which is mainly due to a cellular hyperplasia rather than to increased storage of thyroglobulin
3. Undergo an invariable amelioration of the exophthalmos after the thyroid has been partially excised by surgery
4. Have an increased rate of uptake of plasma iodide into thyroid cells
Ref. p. 911

1154. IN THE MOUNTAINOUS AREAS OF THE WORLD WHERE THERE IS ENDEMIC GOITER, THE PATHOGENETIC FACTORS INCLUDE:
1. The deficit of iodide in the soil and water
2. A deficit in the total amount of thyroid hormones synthesized per day
3. A higher rate of secretion of thyrotropin
4. A hyperplasia of thyroid cells caused by the excess of thyrotropin
Ref. p. 912

SECTION VII - ENDOCRINE PHYSIOLOGY

1155. CONFIRMATORY OF THE DIAGNOSIS OF NON-GENETIC HYPOTHYROIDISM IN A GROWING CHILD ARE A:
1. Short stature
2. Delayed epiphysial growth
3. Low protein bound iodine level in plasma
4. Low iodide uptake   Ref. p. 913

1156. A MYXEDEMATOUS PATIENT WHO RECEIVES A PROPER TREATMENT WILL IN THE INITIAL PHASE OF THIS TREATMENT:
1. Lose an excess of fluid
2. Increase his heart rate
3. Relieve his constipation
4. Have a partial loss of memory   Ref. p. 913

1157. AMONG PROPERTIES OF INSULIN ARE THAT IT:
1. Accelerates the diffusion into muscle cells of non-utilizable sugars which share some structural similarities with glucose, such as galactose molecules
2. Has its effects on glucose transfer, even at 0° C
3. Acts on glucose transfer even if phosphorylation in the cell is blocked
4. Has a very marked accelerating effect of glucose metabolism even in a cell free system (cell membranes broken)
     Ref. p. 916

1158. INSULIN ACCELERATES GLUCOSE TRANSPORT:
1. By the duodenal mucosa
2. By the proximal tubular cells of the renal nephrons
3. Into neurons
4. Into adipose tissue cells   Ref. p. 917

1159. THE TRANSFER OF GLUCOSE INTO CARDIAC MUSCLE FIBERS:
1. Will not take place against a concentration gradient even if insulin is present
2. Occurs by diffusion
3. Is accelerated by an effect of insulin on their membrane
4. Is facilitated by myocardial hypoxia
     Ref. p. 917

1160. IN LIVER CELLS, INSULIN CAUSES:
1. An increased activity of glucokinase
2. A facilitated diffusion through the cell membranes
3. An increased glycogen synthesis after a delay of a few hours
4. Increased gluconeogenesis   Ref. p. 918

1161. HYPOGLYCEMIA INDUCED BY FASTING CAUSES A(N):
1. Increased rate of secretion of insulin
2. Decreased rate of secretion of glucagon
3. Decreased rate of secretion of growth hormone
4. Increased secretion of epinephrine   Ref. p. 919

SECTION VII - ENDOCRINE PHYSIOLOGY

1162. AN INCREASED QUANTITY OF CYCLIC AMP IN THE LIVER:
1. Activates a phosphorylase kinase
2. Is the result of increased concentration of insulin
3. Is the consequence of increased levels of glucagon in postprandial periods
4. Inhibits glucose-6-phosphatase   Ref. p. 919

1163. THE HYPERGLYCEMIA FOUND IN DECOMPENSATED DIABETICS OCCURS SIMULTANEOUSLY WITH A(N):
1. Rise in free fatty acid concentration in blood
2. Rise in acetoacetic acid concentration in blood
3. Lower pH in plasma
4. Increase in plasma osmolarity   Ref. p. 920

1164. INCREASED TURNOVER OF LIPIDS OCCURS IN DIABETES BECAUSE THE LACK OF INSULIN RESULTS IN OR IS CONCOMITANT WITH:
1. A decreased concentration of intracellular glycerophosphate, due to a reduced glycolysis
2. A decreased concentration of cellular TPNH due to a reduced pentose shunt
3. Increased catecholamines and increased cyclic AMP in adipose tissue, which facilitates lipolysis
4. An increased rate of secretion of cortisol, which activates lipolysis
     Ref. p. 920

1165. IT HAS BEEN POSTULATED THAT THE REASON WHY AN INCREASED CONCENTRATION OF FATTY ACIDS IN PLASMA AND IN CELLS WOULD INHIBIT THE METABOLISM OF GLUCOSE, IS BECAUSE WITHIN CELLS THERE IS IN SUCH A CASE A(N):
1. Rise in acetyl Coenzyme A
2. Depression of the conversion of pyruvate to acetyl Coenzyme A
3. Accumulation of citrate
4. Inhibition of phosphofructokinase caused by the excess of citrate
     Ref. p. 921

1166. IN ADDITION TO EFFECTS ON CARBOHYDRATE AND FAT METABOLISM, INSULIN MODIFIES PROTEIN METABOLISM. INSULIN HAS BEEN SHOWN TO:
1. Accelerate the transcellular passage of amino acids
2. Increase protein synthesis in the liver
3. Induce growth in a hypophysectomized and depancreatized animal
4. Inhibit the effect of growth hormone if given concomitantly with the latter   Ref. p. 922

1167. A PATIENT WHO HAS HAD DIABETES MELLITUS FOR 10 YEARS OR MORE, MAY FIND THAT HIS GLYCOSURIA HAS DISAPPEARED EVEN THOUGH HIS BLOOD GLUCOSE LEVELS ARE STILL HIGHER THAN BEFORE. THIS SHOULD SUGGEST THAT HIS:
1. Renal Tm for glucose has increased
2. Renal circulation is largely shunted away from nephrons with low Tm values
3. Distal tubules are now reabsorbing glucose
4. Glomerular filtration rate is severely diminished, while his tubular function is relatively better preserved
     Ref. p. 923

SECTION VII - ENDOCRINE PHYSIOLOGY

1168. DEHYDRATION SEEN IN CASES OF DECOMPENSATED DIABETES IS:
   1. Caused in part by an osmotic diuresis
   2. Limited to the extracellular fluid but does not involve the intracellular fluid
   3. Associated with a severe hyponatremia
   4. Relatively mild   Ref. p. 924

1169. ELECTROLYTES ARE ALTERED AS A RULE IN SEVERE DECOMPEN-SATED DIABETIC ACIDOSIS. THUS, THERE IS IN THIS CONDITION A(N):
   1. Increased concentration of hydrogen ions
   2. Increased concentration of organic ions
   3. Lowering of bicarbonate ion concentration
   4. Decrease in the concentration of chloride ions
                                         Ref. p. 924

1170. IMPORTANT FINDINGS THAT TEND TO RULE OUT THE DIAGNOSIS OF A POSSIBLE CASE OF UNCOMPENSATED DIABETIC ACIDOSIS IN A COMATOSE PATIENT ARE THAT HE:
   1. Had gained weight recently as a result of polyphagia
   2. Has a pH equal to 7.1 in blood, but no ketonemia
   3. Had polyuria and polydypsia but no glucose in the urine
   4. Is in coma but there are no signs of dehydration
                                         Ref. p. 924

1171. IN A HYPERGLYCEMIC PATIENT WHO SHOWS A VERY SLIGHT OR NO HYPOGLYCEMIC RESPONSE TO A TEST DOSE OF INSULIN ADMINIS-TERED INTRAVENOUSLY, THERE IS A POSSIBILITY THAT THE HYPERGLYCEMIA IS DUE TO:
   1. Insufficiency of beta cells of the pancreatic islets, as the only defect
   2. Insufficiency of the alpha cells of the pancreatic islets
   3. An insufficient secretion of cortisol
   4. A relative deficit of insulin together with an excessive secretion of somatotrophin or cortisol   Ref. p. 925

1172. IMPORTANT AIMS IN THE LONG TERM TREATMENT OF DIABETIC PATIENTS ARE:
   1. A careful prevention of glycosuria
   2. The prevention of high insulin concentrations at times when blood glucose may not be controlled by ingestion, as during sleep
   3. The suppression of carbohydrates in the diet
   4. The prevention of hyperlipidemia and hypercholesterolemia by balancing the carbohydrates in the diet with the necessary insulin treatment   Ref. p. 926

1173. AN UTMOST CARE IN THE TREATMENT OF THE COMATOSE DIABETIC PATIENT IS RELATED TO THE NEED TO CORRECT THE BODY POTASSIUM DISTURBANCES. THUS, IT HAS BEEN LEARNED THAT:
   1. Dehydration and acidosis are associated with significant loss of intracellular potassium
   2. Plasma potassium may not be necessarily low in the initial phases of treatment
   3. Hypokalemia may develop rapidly when infusions of saline without potassium are given with insulin and glucose
   4. Death may occur due to acute severe hypokalemia
                                         Ref. p. 926

SECTION VII - ENDOCRINE PHYSIOLOGY

1174. THE ROLE OF GLUCAGON IN CARBOHYDRATE METABOLISM HAS BEEN ELUCIDATED IN PART IN THE RECENT DECADE. IT HAS BEEN LEARNED THAT GLUCAGON:
  1. Causes glycogenolysis in the liver
  2. Causes glycogenolysis in skeletal muscle fibers
  3. Increases levels of cyclic AMP in liver cells
  4. Inhibits gluconeogenesis in liver cells
  Ref. p. 927

1175. IN THE REGULATION OF THE CONCENTRATION OF CALCIUM IN THE EXTRACELLULAR FLUID AND IN PLASMA, THE ABSORPTION OF CALCIUM BY INTESTINAL EPITHELIAL CELLS REPRESENTS A MAJOR REGULATORY MECHANISM. IT IS KNOWN THAT INTESTINAL ABSORPTION OF CALCIUM:
  1. Is increased by a local action of vitamin D on intestinal cells
  2. Is not dependent on parathyroid hormone levels in blood
  3. Is an active process that takes place mainly in the duodenum and jejunum
  4. Occurs for the major part in the colon
  Ref. p. 930

1176. IN A CASE OF MANIFEST TETANY IN A HUMAN:
  1. The concentration of the protein bound calcium fraction must be lowered
  2. The subsequent development of metabolic acidosis is likely to bring about a worsening of the condition
  3. There is no real danger of death if severe hypocalcemia is the only disturbance
  4. Ionic calcium concentration may reach values below 2.3 mEq Liter plasma
  Ref. p. 931

1177. HYPERCALCEMIA MAY BE SUSPECTED IN PATIENTS WHO HAVE:
  1. Hyperreflexia
  2. Euphoria
  3. Broad QRS complexes which last more than 0.15 seconds
  4. Shortened QT interval
  Ref. p. 932

1178. THE CRYSTALLINE STRUCTURE PRESENT IN BONE:
  1. Has one calcium, one magnesium and one potassium atom for each phosphate group
  2. Is that of a hydroxyapatite
  3. Contains chondroitin sulfate as part of the lattice
  4. May adsorb uranium and plutonium very easily
  Ref. p. 932

1179. IF CALCIUM IS SUDDENLY INCREASED IN CONCENTRATION IN BLOOD PLASMA, SUCH AS IN THE CASE OF A MISTAKEN INFUSION, THE MOST LIKELY DEPOSITORY OF THE EXCESS CALCIUM IS:
  1. In the form of an easily soluble salt adsorbed to apatite crystals
  2. In the form of hydroxyapatite lattice
  3. Readily reabsorbable if hypocalcemia were to develop rapidly
  4. In the colon
  Ref. p. 933

SECTION VII - ENDOCRINE PHYSIOLOGY

1180. OSTEOBLASTS FORM NEW BONE:
1. Until a very narrow Haversian canal is all that is left from a previously large space
2. Secreting alkaline phosphatase
3. By deposition of a collagenous matrix on which the apatite crystal is formed
4. When stimulated by parathyroid hormone
Ref. pp. 933, 934

1181. AN ELEVATED ALKALINE PHOSPHATASE SHOULD BE EXPECTED WHEN BLOOD IS OBTAINED FROM:
1. Growing children
2. Adults with osteomalacia
3. Adults shortly after a bone fracture
4. Rachitic children
Ref. p. 935

1182. PARATHYROID HORMONE IS KNOWN TO:
1. Increase the rate of tubular reabsorption of filtered calcium in the proximal tubule of renal nephrons
2. Decrease the rate of tubular reabsorption of ionic phosphate by proximal tubular cells
3. Increase the formation of osteoclasts
4. Cause the release of calcium phosphate from bone crystals
Ref. pp. 936, 937

1183. A FUNCTIONAL HYPERPLASIA OF PARATHYROID GLANDS SHOULD BE EXPECTED TO OCCUR:
1. When there is chronic deficit in vitamin D in the diet
2. In children who have a diet rich in milk and cheese
3. In lactating mothers
4. In vitamin D intoxication
Ref. p. 937

1184. A PATIENT WHO LOSES LARGE VOLUMES OF INTESTINAL CONTENTS PER DAY DUE TO FREQUENT BOWEL MOVEMENTS MAY LOSE A LARGE AMOUNT OF CALCIUM IN A SHORT TIME. THIS LOSS ELICITS SEVERAL COMPENSATORY REACTIONS AMONG WHICH WE SHOULD INCLUDE:
1. A very rapid flux of exchangeable calcium adsorbed to bone crystals
2. Secretion of parathyroid hormone that becomes increased in rate even before the flux of exchangeable calcium into the extracellular fluid is increased
3. A gradual effect of parathyroid hormone on renal reabsorption of calcium
4. An almost immediate effect of parathyroid hormone on apatite crystals, as soon as hypocalcemia is first developed
Ref. pp. 936-939

1185. THE REGULATION OF CALCIUM HOMEOSTASIS IS CONTROLLED MAINLY BY PARATHYROID HORMONE, VITAMIN D AND CALCITONIN. THE LATTER:
1. Acts by promoting calcium reabsorption from bones
2. Acts very rapidly and after calcium ion concentration has been increased for a short time
3. Seems to be secreted by the parathyroid and not by the thyroid
4. Is secreted at very low rates, if at all, when there is hypocalcemia
Ref. p. 939

SECTION VII - ENDOCRINE PHYSIOLOGY

1186. A PATIENT WHO IN THE PROCESS OF THE SURGICAL ABLATION OF PART OF HIS THYROID GLAND LOSES ALL OF HIS PARATHYROIDS WILL PRESENT THE FOLLOWING SYMPTOMS AND SIGNS AFTER A SHORT PERIOD OF TIME (FEW HOURS):
   1. His blood calcium concentration will drop below 7 mg per 100 ml plasma
   2. His blood phosphate concentration will rise to about 12 mg per 100 ml plasma
   3. Tremor and muscle spasm will make motion and even breathing very difficult
   4. Phosphaturia will be massive   Ref. p. 939

1187. THE BEST TREATMENT FOR CHRONIC HYPOPARATHYROIDISM INCLUDES:
   1. Parathyroid hormone injections 3 times per day
   2. Very low phosphate in the diet
   3. High calcium diet and parathyroid hormone injections every week
   4. Vitamin D and a high calcium diet   Ref. p. 939

1188. ONE SHOULD BE ALERT TO EXPLORE FURTHER TO RULE OUT AN EXCESSIVELY HIGH RATE OF SECRETION OF THE PARATHYROID GLANDS IF A PATIENT HAS:
   1. A bone fracture appearing after a minimal trauma
   2. One or numerous bone cysts detected by X-rays
   3. Frequent occurrence of renal calculi
   4. A large goiter   Ref. p. 940

1189. IN THE PATHOGENESIS OF RICKETS, THE LACK OF VITAMIN D IN THE DIET CAUSES:
   1. A significant lowering of the concentration of phosphate ions
   2. Hyperplasia of the parathyroid glands
   3. A barely diminished calcium ion concentration
   4. An osteoid-rich, mineral-poor metaphysial bone
       Ref. p. 941

1190. CARIOUS TEETH SHOULD BE EXPECTED TO OCCUR MORE OFTEN:
   1. In teeth formed in areas where no fluoride is added to the diet during infancy and childhood
   2. In persons who have hyperparathyroidism
   3. Presumably in teeth exposed to prolonged contact with acid producing bacteria
   4. In women who have had numerous pregnancies and whose diet was poor in calcium   Ref. p. 943

1191. SPERMATOCYTES IN THE HUMAN TESTIS:
   1. Undergo a first meiotic division when a primary spermatocyte divides to form a secondary spermatocyte
   2. Split their X-Y pair during the division of a secondary spermatocyte to form a spermatid
   3. Secrete testosterone
   4. Are stimulated to divide under the influence of pituitary FSH
       Ref. pp. 946, 954

SECTION VII - ENDOCRINE PHYSIOLOGY

1192. ACTIVE MOTILITY OF SPERMATOZOA IS AN ESSENTIAL FEATURE OF FERTILE SPERM. THIS MOTILITY IS:
1. Acquired in the seminiferous tubules
2. Increased in neutral or alkaline pH
3. Stimulated by vaginal secretions
4. Depressed by an acid pH   Ref. p. 947

1193. A HUMAN MALE MAY BECOME INFERTILE IF:
1. His seminiferous tubules are constantly exposed to 38º C
2. His semen contains less than 35-40 million spermatozoa per ml
3. Fifty per cent of his spermatozoa exhibit little or no motility
4. His semen is retained for more than 3 days in the female's internal sexual organs without contact with an ovum
                                          Ref. p. 948

1194. INTERSTITIAL CELLS OF THE HUMAN TESTIS, WHICH ARE BELIEVED TO BE THE SOURCE OF TESTOSTERONE, ARE:
1. Abundant in the postpubertal male (age 14-60)
2. As abundant in the young male (5-10 years) as in the adult male
3. Numerous in the newborn
4. Non existent in the cryptorchidic male
                                          Ref. p. 951

1195. THE SO-CALLED 17-KETOSTEROIDS:
1. Are inert metabolic products of testosterone, and have no physiological action
2. Originate exclusively in the testis
3. Are absent in the normal adult female
4. May be abnormally elevated in a functional adenoma of the adrenal
                                          Ref. p. 951

1196. TESTOSTERONE, THE MAIN ANDROGENIC STEROID IN MAN:
1. Is absent in the embryo, during the whole duration of fetal life
2. Can induce testicular descent in the cryptorchidic male
3. Causes increased catabolism of connective tissue that enters in the composition of bone
4. May contribute to short stature in sexually precocious males
                                          Ref. p. 953

1197. THE TESTES ARE UNDER THE CONTROL OF THE PITUITARY AND IT IS PRESENTLY BELIEVED THAT:
1. FSH induces the conversion of the primary spermatocyte into secondary spermatocytes
2. LH induces the secretion of testosterone from the interstitial cells
3. An excess of testosterone decreases the rate of secretion of LH producing cells in the pituitary
4. The pituitary of a male is different in its gonadotrophic potential from that of a female   Ref. pp. 954, 955

SECTION VII - ENDOCRINE PHYSIOLOGY 223

1198. A MALE EUNUCH HAS SEVERAL UNIQUE SOMATIC FEATURES WHICH ARE NOT FOUND IN THE NORMAL ADULT MALE. THESE FEATURES DESCRIBE VIVIDLY AND IN A SORT OF A NEGATIVE WAY THE ANDROGENIC EFFECTS OF TESTOSTERONE. THUS AN ADULT EUNUCH HAS:
1. A relatively weak musculature
2. Relatively thinner and weaker bones
3. An infantile penis
4. A relatively tall height when compared to other male members of his family Ref. p. 956

1199. THE PROCESS OF OVULATION IS PROBABLY THE CENTRAL EVENT IN THE SEQUENCE OF PHENOMENA TAKING PLACE IN CYCLIC FASHION IN THE FEMALE REPRODUCTIVE SYSTEM. IN ORDER THAT OVULATION MAY OCCUR, IT IS INDISPENSABLE THAT THE:
1. Pituitary secretes FSH at increasing rates for about 10-15 days in the human
2. Pituitary secretes FSH and LH at the same rate throughout the preparatory period
3. Ventromedial nuclei of the hypothalamus discharge more of their LH releasing factor at the end of the preparatory period
4. Concentration of estrogens and progesterone in blood perfusing the hypothalamus be at its highest peak in the last 24 hours before ovulation should take place Ref. pp. 960, 965, 966

1200. AFTER OVULATION, THERE IS A PERIOD OF TIME WHICH LASTS ABOUT 7-10 DAYS IN THE HUMAN, WHEN ANOTHER OVULATION IS IMPOSSIBLE. THIS FACT IS THE KEY REASON WHY MOST OF THE ANTICONCEPTIONAL DRUGS EXERT THEIR EFFECT. THUS, AFTER OVULATION, AND FOR A CERTAIN PERIOD OF TIME AFTERWARD:
1. The granulosa cells of the ruptured follicle are transformed into luteinized cells
2. Luteinized cells and probably theca lutein cells secrete greater quantities of estrogen and progesterone
3. High titers of estrogen and progesterone probably inhibit the hypothalamic centers that form FSH and LH releasing factors
4. The ovary itself does not have space for another growing follicle Ref. pp. 965, 966

1201. THE RATHER HIGH CONCENTRATION OF ESTROGENS IN BLOOD FOUND CYCLICALLY IN ADULT HUMAN FEMALES CAUSES:
1. Secretory changes to appear in the endometrial glands
2. An increase in ciliated cells in the fallopian tubes
3. Marked growth of secretory acini in female breasts prior to lactation
4. Increased deposition of calcium and phosphate in the bone matrix Ref. p. 963

1202. IN WOMEN, WITHIN THE FIRST FEW MONTHS AFTER MENOPAUSE, THE:
1. Blood levels of progesterone drops to zero
2. Rate of secretion of pituitary FSH is increased several fold above premenopausal levels
3. Hypothalamic pituitary system can still be inhibited by exogenous estrogens
4. Endometrium becomes unresponsive to exogenous estrogens Ref. p. 968

SECTION VII - ENDOCRINE PHYSIOLOGY

1203. THE FOLLOWING OBSERVATIONS HAVE BEEN MADE IN THE HUMAN FEMALE IN RELATION TO MENSTRUAL CYCLES:
1. The only epithelial cells that remain in the endometrium after a menstrual hemorrhage has been completed are those present at the bottom of the endometrial glands
2. Tortuosity of endometrial vessels is maximal during the secretory phase
3. Estrogens and progesterone in blood fall to their lowest concentration a few hours before the hemorrhagic phase
4. Vascular spasm of endometrial arterioles is maximal a few hours before the hemorrhagic phase   Ref. p. 969

1204. A GRANULOSA CELL TUMOR OF THE OVARY USUALLY SECRETES ESTROGENS AT AN EXCESSIVELY HIGH RATE. THIS MAY LEAD, IN AN ADULT WOMAN, TO:
1. Endometriosis
2. A high rate of secretion of FSH
3. A high rate of secretion of LH
4. Endometrial hemorrhage   Ref. p. 970

1205. IT IS TYPICAL OF PATIENTS WHO HAVE ENDOMETRIOSIS TO COMPLAIN OF:
1. Constipation
2. Infertility
3. Abdominal pain concident with menstrual periods
4. Multiple pregnancies   Ref. p. 970

1206. IN THE MATURATION PROCESS THAT LEADS TO THE FORMATION OF A FERTILIZABLE OVUM IN THE HUMAN FEMALE:
1. The meiotic division occurs at the time when the secondary oocyte gives rise to the second polar body
2. The mature ovum has a Y chromosome and 22 autosomes
3. Numerous ova may grow simultaneously to almost full-size in each cycle
4. The rule is that several ova are actually ejected from the ovary in each ovulation   Ref. p. 975

1207. PLACENTAL NUTRITION OF THE HUMAN EMBRYO IS ARRANGED SO THAT:
1. Blood contained in the vessels of the umbilical cord is more oxygenated in the umbilical veins than in the umbilical arteries
2. The cytotrophoblast may be prominently seen histologically until term
3. Permeability of the placental layers to nutrients present in maternal blood is presumably lower in the first 4 months than in the last 4 months of gestation
4. The oxygen dissociation curve for fetal hemoglobin is "shifted to the left" of that of adult hemoglobin   Ref. pp. 977, 978

1208. THE INITIATION OF PREGNANCY AFTER SUCCESSFUL FERTILIZATION OF AN OVUM BY A SPERMATOZOON REQUIRES THAT:
1. Chorionic gonadotrophin be secreted by trophoblastic cells
2. Chorionic gonadotrophin stimulate the corpus luteum cells
3. The corpus luteum secrete estrogens and progesterone at a higher rate
4. The pituitary secrete increased quantities of LH
     Ref. pp. 979, 980

SECTION VII - ENDOCRINE PHYSIOLOGY 225

1209. HUMAN PLACENTAL LACTOGEN IS KNOWN TO:
1. Be secreted within the first month of pregnancy but not afterwards
2. Have an insulin like effect on the tissues of the embryo
3. Inhibit cellular growth when it is secreted at a high rate
4. Have a great functional similarity with pituitary somatotrophin
Ref. p. 980

1210. DURING PREGNANCY IN THE HUMAN SPECIES:
1. The bulk of estrogens and progesterone that circulate in blood during the third trimester is secreted by the placenta
2. Secretion of pituitary FSH and LH are greatly increased in the first trimester of pregnancy
3. The rate of release of thyroid hormone from the thyroid gland is increased
4. The secretion of cortisol is inhibited
Ref. p. 981

1211. HORMONES CAUSING AN EXPANSION OF THE EXTRACELLULAR FLUID VOLUME DURING PREGNANCY ARE:
1. Estrogens
2. Progesterone
3. Cortisol
4. Aldosterone   Ref. pp. 982, 981

1212. IN PREGNANCY, A POSITIVE CORRELATION EXISTS BETWEEN:
1. Nausea and vomiting seen during pregnancy on one side and secretion rates of chorionic gonadotropin on the other
2. Reduction in glomerular filtration rate and toxemia
3. Reduction in cerebral blood flow and eclampsia
4. The ratio of progesterone in blood and increased uterine contractility
    estrogen
Ref. p. 984

1213. THE FOLLOWING EVENTS APPLY TO A NORMAL PARTURITION IN THE HUMAN:
1. Resistance to blood flow in the uterine wall and to placental circulation increases significantly while the uterus contracts
2. Contraction pains could be suppressed if the pudendal nerves (which enter the sacral portion of the spinal cord) were locally anesthetized
3. The dilatation of the cervical canal requires a much longer time than the process of the expulsion of the fetus
4. Placental separation is usually associated with 1 liter of blood loss
Ref. pp. 986, 987

1214. THE COMPLEX PROCESS OF PROTEIN, CARBOHYDRATE AND FAT SYNTHESIS UNDERLYING THE SECRETION OF MILK REQUIRES A HIGH HORMONAL SECRETORY RATE OF:
1. Prolactin
2. Estrogens
3. Oxytocin
4. Progesterone   Ref. p. 988

SECTION VII - ENDOCRINE PHYSIOLOGY

1215. BEFORE LACTATION CAN OCCUR IN THE HUMAN FEMALE, THE FOLLOWING ENDOCRINE EVENTS MUST TAKE PLACE:
1. The hypothalamic prolactin-inhibitory factor must be supressed
2. The rate of secretion of estrogens must increase
3. The rate of secretion of progesterone must decrease
4. The supraoptic and paraventricular nuclei must be inhibited
Ref. p. 989

1216. WHEN COW'S MILK AND HUMAN MILK ARE COMPARED IN THEIR FRACTIONAL CONCENTRATIONS, ONE FINDS THAT THERE IS IN COW'S MILK THREE TIMES GREATER CONCENTRATION OF:
1. Glucose
2. Lactalbumin
3. Fat
4. Casein
Ref. Table 82-1, p. 989

1217. ON A WEIGHT BASIS THE PROPORTION OF TRANSFER OF NUTRIENTS FROM LACTATING MOTHER TO SUCKLING INFANT IS GREATEST FOR:
1. Calcium
2. Phosphate
3. Fats
4. Lactose
Ref. p. 990

1218. IT HAS BEEN DETERMINED THROUGH EMBRYOLOGICAL STUDIES IN THE HUMAN THAT:
1. The embryo's heart starts contracting about 4 weeks after fertilization
2. Myelinization of the corticospinal tract is complete by the 6th month of gestation
3. The bone marrow starts forming blood cells at about the 12th week after fertilization
4. Glomerular filtration does not start until after delivery
Ref. p. 992

1219. IN THE NEWBORN INFANT THE WORK OF BREATHING, CALCULATED FROM PRESSURE-VOLUME CURVES:
1. Is equal in magnitude for the first twenty breathing movements
2. Is much smaller in the first breathing than in subsequent respiratory movements done an hour later
3. Overcomes at first only the viscosity of the fluid within alveoli, but does not oppose elastic forces
4. Decreases rapidly in the first hour of breathing
Ref. Fig. 83-3, p. 994

1220. MOST LIKELY CANDIDATES TO DEVELOP HYPOXIA IN THE SO-CALLED HYALINE MEMBRANE DISEASE ARE FETUSES:
1. Of older mothers
2. Delivered by forceps
3. Whose mothers are hypoglycemic
4. Who are born prematurely
Ref. p. 994

SECTION VII - ENDOCRINE PHYSIOLOGY

1221. THE PATH OF LEAST RESISTANCE FOR BLOOD FLOWING THROUGH THE FETAL CIRCULATORY SYSTEM CONDUCTS THIS BLOOD IN SUCH A MANNER THAT THE:
1. Fetal liver is only slightly perfused by blood entering through the umbilical cord
2. Lungs are largely bypassed
3. Left ventricle has a greater work load than the right ventricle
4. Blood flow through the ductus arteriosus is mainly concerned with the perfusion of the lungs so that its direction preferentially is from aorta to pulmonary artery Ref. pp. 994, 995

1222. CARDIOVASCULAR FUNCTION IN THE NEONATE IS DIFFERENT FROM THAT IN THE EMBRYO, IN THAT, IN THE NEONATE:
1. The pulmonary vascular resistance is lower than in the embryo
2. The systemic vascular resistance is lower than in the embryo
3. The left atrial pressure becomes higher than the right atrial pressure
4. Cardiac output per unit of body mass is greater than in the embryo
Ref. p. 996

1223. EXCESSIVE HEMOLYSIS MAY OCCUR IN THE NEONATE IN CASES:
1. Where the main difficulty is one of immaturity of the enzymatic systems responsible for conjugation of bilirubin within the liver cells
2. Of vitamin K deficiency
3. Of physiologic anemia
4. Of erythroblastosis fetalis Ref. p. 997

1224. THERE ARE SEVERAL PHYSIOLOGICAL ADAPTATIONS IN THE INITIAL FEW HOURS THAT FOLLOW IMMEDIATELY AFTER BIRTH. THUS IN THE NEONATE THERE IS NORMALLY A:
1. Tendency of the body temperature to stay about 2^0 C above the normal
2. Cessation of flow through the ductus venosus
3. Urinary osmolarity ratio that may be as high as 6.0
   Plasma
4. Relative deficiency in pancreatic amylase
Ref. p. 998

1225. A PREMATURE INFANT SHOULD NOT:
1. Be exposed to 100% oxygen for a prolonged period of time
2. Be maintained in an open room at an ambient temperature of 20° C
3. Have a fat rich diet
4. Have a diet rich in calcium and vitamin D
Ref. p. 1000

ANSWER KEY

The author has taken great pains to check thoroughly the questions and answers. However, in a volume of this size, some ambiguities and possible inaccuracies may appear. Therefore, if in doubt, consult your references.

THE PUBLISHERS

SECTION I

1. C	48. E	95. C	142. A	189. E
2. C	49. C	96. B	143. D	190. B
3. A	50. C	97. A	144. D	191. E
4. C	51. C	98. E	145. D	192. D
5. A	52. E	99. C	146. A	193. B
6. A	53. B	100. C	147. B	194. A
7. D	54. E	101. C	148. D	195. B
8. B	55. A	102. B	149. D	196. C
9. B	56. A	103. D	150. A	197. E
10. A	57. E	104. D	151. C	198. C
11. B	58. D	105. D	152. C	199. C
12. A	59. E	106. D	153. E	200. C
13. C	60. C	107. E	154. E	201. A
14. B	61. B	108. D	155. C	202. D
15. C	62. C	109. D	156. D	203. E
16. C	63. A	110. A	157. D	204. D
17. C	64. E	111. E	158. D	205. D
18. A	65. C	112. B	159. D	206. B
19. B	66. B	113. D	160. B	207. C
20. D	67. A	114. A	161. D	208. B
21. A	68. C	115. C	162. D	209. C
22. B	69. A	116. B	163. E	210. B
23. C	70. B	117. C	164. A	211. A
24. C	71. A	118. B	165. C	212. B
25. A	72. B	119. D	166. A	213. B
26. C	73. E	120. E	167. E	214. C
27. D	74. D	121. C	168. D	215. D
28. A	75. B	122. B	169. A	216. C
29. A	76. A	123. B	170. D	217. B
30. C	77. A	124. C	171. E	218. A
31. B	78. A	125. B	172. C	219. C
32. D	79. A	126. C	173. A	220. B
33. E	80. D	127. C	174. C	221. B
34. D	81. A	128. B	175. D	222. E
35. B	82. E	129. B	176. B	223. D
36. D	83. A	130. E	177. B	224. B
37. B	84. C	131. B	178. B	225. C
38. C	85. B	132. A	179. D	226. A
39. A	86. D	133. B	180. D	227. D
40. E	87. A	134. D	181. B	228. C
41. E	88. D	135. C	182. C	229. B
42. F	89. C	136. E	183. B	230. A
43. B	90. B	137. D	184. E	231. A
44. C	91. A	138. D	185. B	232. C
45. D	92. B	139. D	186. A	233. B
46. A	93. C	140. E	187. B	234. A
47. D	94. E	141. C	188. B	235. C

ANSWER KEY

236. C	288. C	343. B	398. A	450. D
237. B	289. D	344. D	399. C	451. D
238. B	290. B	345. B	400. A	452. C
239. C	291. C	346. C	401. B	453. C
240. C	292. B	347. A	402. B	454. B
241. C	293. B	348. C	403. B	455. B
242. C	294. B	349. C	404. D	456. D
243. B	295. D	350. A	405. B	457. B
244. A	296. C	351. B	406. A	458. C
245. B	297. D	352. B	407. B	459. B
246. B	298. E	353. A	408. A	460. A
247. E	299. B	354. C	409. A	461. E
248. D	300. B	355. C	410. C	462. B
249. B	301. E	356. B	411. D	463. B
250. C	302. C	357. D	412. C	464. C
251. D	303. D	358. A	413. A	465. C
252. A	304. C	359. A	414. B	466. B
253. B	305. D	360. B	415. A	467. A
254. A	306. E	361. B	416. B	468. A
255. B	307. C	362. A	417. D	469. C
256. B	308. A	363. B		470. D
257. C	309. E	364. B	**SECTION III**	471. C
258. C	310. B	365. B		472. D
259. C	311. D	366. C	418. D	473. B
260. C	312. E	367. A	419. B	474. E
261. D	313. A	368. A	420. B	475. C
262. C	314. B	369. D	421. C	476. C
263. C	315. E	370. C	422. B	477. D
264. C	316. B	371. A	423. B	478. D
265. A	317. D	372. A	424. D	479. E
266. D	318. A	373. D	425. A	480. B
267. D	319. D	374. D	426. B	481. B
268. C	320. A	375. D	427. B	482. A
	321. B	376. A	428. B	483. B
SECTION II	322. C	377. A	429. C	484. D
	323. D	378. A	430. B	485. C
269. D	324. A	379. E	431. D	486. C
270. B	325. B	380. C	432. D	487. A
271. B	326. D	381. A	433. B	488. B
272. E	327. C	382. A	434. B	489. D
273. B	328. A	383. A	435. C	490. D
274. D	329. C	384. A	436. C	491. B
275. A	330. C	385. C	437. C	492. B
276. B	331. A	386. D	438. B	493. C
277. B	332. C	387. B	439. D	494. A
278. A	333. D	388. C	440. D	495. C
279. E	334. A	389. C	441. B	496. A
280. C	335. A	390. A	442. B	497. B
281. A	336. C	391. D	443. D	498. A
282. B	337. C	392. D	444. C	499. B
283. A	338. A	393. A	445. B	500. B
284. B	339. D	394. D	446. C	501. A
285. B	340. E	395. D	447. D	502. A
286. C	341. A	396. C	448. B	503. D
287. C	342. B	397. C	449. C	504. B

ANSWER KEY

505. C	557. E	612. A	667. A	722. D
506. B	558. E	613. D	668. C	723. C
507. E	559. D	614. D	669. E	724. B
508. B	560. A	615. C	670. B	725. E
509. D	561. E	616. D	671. A	726. C
510. C	562. E	617. D	672. D	727. D
511. D	563. D	618. D	673. A	728. B
512. C	564. E	619. D	674. B	729. D
513. B	565. B	620. C	675. C	730. A
514. C	566. C	621. D	676. D	731. E
515. A	567. A	622. B	677. C	732. D
516. A	568. E	623. D	678. C	733. D
517. E	569. C	624. D	679. C	734. E
518. E	570. A	625. E	680. D	735. B
519. A	571. A	626. C	681. D	736. A
520. C	572. B	627. D	682. B	737. D
521. C	573. C	628. C	683. C	738. A
522. A	574. D	629. D	684. B	739. B
523. A	575. A	630. A	685. A	740. E
524. A	576. C	631. C	686. C	741. D
525. A	577. E	632. D	687. E	742. A
526. C	578. D	633. B	688. C	743. C
527. A	579. A	634. B	689. D	744. C
528. D	580. D	635. C	690. D	745. C
529. C	581. A	636. D	691. C	746. C
530. B	582. B	637. E	692. C	747. B
531. D	583. A	638. D	693. B	748. A
532. E	584. A	639. C	694. D	749. C
533. C	585. A	640. E	695. C	750. A
534. D	586. A	641. A	696. A	751. B
535. D	587. C	642. B	697. C	752. B
536. C	588. C	643. C	698. C	753. D
537. B	589. C	644. E	699. B	754. D
538. E	590. C	645. A	700. B	755. A
539. D	591. A	646. D	701. D	756. B
540. A	592. A	647. C	702. B	757. A
541. E	593. A	648. D	703. B	758. A
542. D	594. A	649. D	704. A	759. B
543. D	595. B	650. D	705. A	760. A
544. B	596. A	651. D	706. D	761. A
545. B	597. C	652. C	707. A	762. B
	598. C	653. E	708. C	763. E
SECTION IV	599. A	654. C	709. B	764. C
	600. D	655. B	710. B	765. A
546. E	601. C	656. A	711. D	766. D
547. E	602. A	657. D	712. D	767. D
548. B	603. B	658. A	713. D	768. D
549. C	604. A	659. B	714. D	769. D
550. E	605. C	660. A	715. E	770. C
551. C	606. A	661. B	716. D	771. A
552. E	607. B	662. C	717. A	772. C
553. B	608. D	663. C	718. B	773. A
554. D	609. D	664. A	719. A	774. B
555. A	610. C	665. A	720. D	775. E
556. A	611. D	666. B	721. C	776. D

ANSWER KEY

777. D	832. C	887. D	939. A	991. B
778. C	833. A	888. D	940. B	992. C
779. D	834. D	889. C	941. A	993. A
780. C	835. D	890. D	942. C	994. B
781. A	836. B	891. E	943. D	995. C
782. B	837. D	892. D	944. B	996. D
783. E	838. C	893. A	945. A	997. D
784. C	839. A	894. D	946. D	998. E
785. C	840. A	895. C	947. B	999. E
786. A	841. A	896. D	948. B	1000. D
787. A	842. C	897. D	949. A	1001. C
788. E	843. C	898. E	950. C	1002. D
789. A	844. C	899. C	951. C	1003. A
790. A	845. D	900. A	952. C	1004. C
791. B	846. B	901. A	953. B	1005. C
792. D	847. E	902. D	954. A	1006. E
793. D	848. E	903. A	955. A	1007. E
794. B	849. C	904. B	956. C	1008. A
795. E	850. C	905. D	957. E	1009. B
796. D	851. D	906. E	958. B	1010. D
797. C	852. E	907. B	959. C	1011. A
798. C	853. D	908. D	960. B	1012. C
799. D	854. C	909. C	961. A	1013. E
800. D	855. E	910. E	962. A	1014. D
801. D	856. A	911. D	963. A	1015. A
802. A	857. D	912. E	964. B	1016. B
803. A	858. B		965. E	1017. B
804. D	859. E	SECTION V	966. D	1018. A
805. E	860. A		967. C	1019. B
806. E	861. D	913. D	968. B	1020. C
807. D	862. D	914. D	969. A	1021. D
808. B	863. C	915. C	970. C	1022. D
809. A	864. D	916. B	971. D	1023. A
810. D	865. B	917. E	972. B	1024. C
811. B	866. C	918. B	973. D	1025. D
812. A	867. C	919. A	974. C	1026. E
813. C	868. D	920. C	975. B	1027. E
814. E	869. A	921. E	976. A	1028. C
815. E	870. C	922. D	977. B	1029. C
816. D	871. B	923. E		1030. C
817. C	872. B	924. A	SECTION VI	1031. D
818. A	873. A	925. C		1032. E
819. D	874. E	926. A	978. D	1033. D
820. C	875. C	927. B	979. E	1034. B
821. D	876. A	928. C	980. A	1035. D
822. A	877. A	929. A	981. C	1036. B
823. D	878. E	930. C	982. D	1037. D
824. D	879. A	931. C	983. E	1038. E
825. C	880. A	932. C	984. D	1039. E
826. A	881. C	933. B	985. A	1040. A
827. B	882. E	934. B	986. A	1041. A
828. A	883. A	935. A	987. D	1042. C
829. E	884. C	936. B	988. D	1043. C
830. E	885. D	937. B	989. E	1044. C
831. C	886. D	938. D	990. E	1045. A

ANSWER KEY

1046. A	1101. A	1153. C	1208. A
1047. C	1102. E	1154. E	1209. D
1048. D	1103. A	1155. A	1210. B
1049. D	1104. D	1156. A	1211. E
1050. D	1105. B	1157. A	1212. A
1051. C	1106. D	1158. D	1213. B
1052. C	1107. A	1159. E	1214. B
1053. A	1108. B	1160. B	1215. B
1054. D	1109. A	1161. D	1216. D
1055. A	1110. D	1162. B	1217. D
1056. A	1111. C	1163. E	1218. B
1057. B	1112. D	1164. E	1219. D
1058. C	1113. C	1165. E	1220. D
1059. E	1114. A	1166. A	1221. A
1060. A	1115. A	1167. D	1222. B
1061. A		1168. B	1223. D
1062. A	**SECTION VII**	1169. E	1224. C
1063. A		1170. E	1225. A
1064. E	1116. A	1171. D	
1065. C	1117. C	1172. C	
1066. A	1118. B	1173. E	
1067. A	1119. E	1174. B	
1068. D	1120. A	1175. B	
1069. C	1121. B	1176. D	
1070. C	1122. E	1177. D	
1071. A	1123. D	1178. C	
1072. C	1124. C	1179. B	
1073. D	1125. C	1180. A	
1074. B	1126. E	1181. E	
1075. E	1127. B	1182. E	
1076. B	1128. C	1183. B	
1077. D	1129. E	1184. B	
1078. C	1130. A	1185. C	
1079. D	1131. C	1186. A	
1080. D	1132. E	1187. D	
1081. C	1133. D	1188. A	
1082. B	1134. B	1189. E	
1083. D	1135. E	1190. B	
1084. A	1136. A	1191. C	
1085. E	1137. B	1192. C	
1086. A	1138. E	1193. E	
1087. C	1139. D	1194. B	
1088. E	1140. B	1195. D	
1089. B	1141. C	1196. C	
1090. A	1142. A	1197. A	
1091. B	1143. E	1198. E	
1092. A	1144. D	1199. B	
1093. C	1145. E	1200. A	
1094. D	1146. A	1201. C	
1095. D	1147. A	1202. A	
1096. C	1148. D	1203. E	
1097. B	1149. C	1204. D	
1098. C	1150. A	1205. A	
1099. C	1151. B	1206. B	
1100. A	1152. B	1207. B	

**AVAILABLE AT YOUR LOCAL BOOKSTORE
OR USE THIS ORDER FORM**

MEDICAL EXAMINATION PUBLISHING CO., INC.
65-36 Fresh Meadow Lane, Flushing, N.Y. 11365

Date: _____

Please send me the following books:

_____

_____

_____

_____

_____

☐ Payment enclosed to save postage.

☐ Bill me. I will remit payment within 30 days.

Name _____

Address _____

City & State _____ Zip _____

(Please print)

D. A. S.

**AVAILABLE AT YOUR LOCAL BOOKSTORE
OR USE THIS ORDER FORM**

MEDICAL EXAMINATION PUBLISHING CO., INC.
65-36 Fresh Meadow Lane, Flushing, N.Y. 11365

Date: _____

Please send me the following books:

_____

_____

_____

_____

_____

☐ Payment enclosed to save postage.

☐ Bill me. I will remit payment within 30 days.

Name _____

Address _____

City & State _____ Zip _____

(Please print)

OTHER BOOKS AVAILABLE

ITEMS	Code	Unit Price	ITEMS	Code	Unit Price
MEDICAL OUTLINE SERIES *(Cont'd.)*			Institutional Laundry Journal Articles	789	$8.00
Endocrinology	614	$10.00	Lithium & Psychiatry Journal Articles	520	15.00
Histology	662	8.00	Psychosomatic Medicine Current J. Art.	788	12.00
Otolaryngology	661	8.00	Outpatient Services Journal Articles, 2nd Ed.	797	10.00
Psychiatry	621	8.00	Outpatient Services Journal Articles, 1st Ed.	794	8.00
Urology	611	8.00	Selected Papers in Inhalation Therapy	523	10.00
SELF-ASSESSMENT BOOKS			**TYPIST HANDBOOKS**		
Self-Assess. Cur. Knldge - Biochemistry	266	7.50	Medical Typist's Guide for Hx & Phys.	976	4.50
S.A.C.K. in Cardiovascular Diseases	275	10.00	Radiology Typist Handbook	981	4.50
S.A.C.K. in Diagnostic Radiology	278	10.00	Surgical Typist Handbook	991	4.50
S.A.C.K. in Family Practice	261	10.00	Transcribers Guide to Med. Terminology	973	4.50
S.A.C.K. in Infectious Diseases	263	10.00	**ESSAY Q. & A. REVIEW BOOKS**		
S.A.C.K. in Internal Medicine	257	10.00	Blood Banking Principles Rev.	339	8.00
S.A.C.K. in Neurology	254	10.00	Cardiology Review	337	10.00
S.A.C.K. for Nurse Anesthetist	715	7.50	Colon & Rectal Surg. Cont. Ed. Rev.	338	10.00
S.A.C.K. in Obstet./Gynecology	260	10.00	Neurology Review	345	10.00
S.A.C.K. in O.R. Techn.	474	7.50	Obstet. Nursing Cont. Ed. Rev.	350	5.00
S.A.C.K. in Otolaryngology	270	10.00	Ophthalmology Review	347	10.00
S.A.C.K. in Pathology	253	10.00	Orthopedics Review	349	10.00
S.A.C.K. in Pediatrics	256	10.00	Psychiatry Cont. Ed. Rev.	352	10.00
S.A.C.K. in Psychiatry	252	10.00	Psych./Mental Hlth. Nursing Cont. Ed. Rev.	351	5.00
S.A.C.K. in Pulmonary Diseases	271	10.00	**OTHER BOOKS**		
S.A.C.K. in Rheumatology	258	10.00	Acid Base Homeostasis	601	4.00
S.A.C.K. in Surgery	250	10.00	Allergy Annual Review	325	12.00
S.A.C.K. in Surgery for Family Physicians	259	10.00	Bailey & Love's Short Practice of Surgery	900	20.00
S.A.C.K. in Urology	251	10.00	Benign & Malignant Bladder Tumors	932	15.00
S.A.C.K. in X-Ray Tech.	274	7.50	Blood Groups	860	2.50
MEDICAL HANDBOOKS			Clinical Diagnostic Pearls	730	4.50
E.N.T. Emergencies	639	8.00	Concentrations of Solutions	602	3.00
Medical Emergencies	635	8.00	Critical Care Manual	983	10.00
Neurology	604	8.00	Cryogenics in Surgery	754	24.00
Obstetrical Emergencies	634	8.00	Diagnosis & Treatment of Breast Lesions	748	15.00
Ophthalmologic Emergencies	633	8.00	Emergency Care Manual	984	7.50
Pediatric Anesthesia	637	8.00	English-Spanish Guide for Med. Personnel	721	2.50
Pediatric Neurology	636	10.00	Fundamental Orthopedics	603	4.50
PRACTICAL'POINTS BOOKS			Guide to Medical Reports	962	4.50
In Anesthesiology	700	10.00	Human Anatomical Terminology	982	3.00
In Gastroenterology	733	7.00	Illustrated Laboratory Techniques	919	10.00
In Pediatrics	702	10.00	Introduction to Acupuncture	753	5.00
PRACTITIONERS GUIDES			Introduction to Blood Banking	975	8.00
OB-Gynecology Disorders	704	10.00	Introduction to the Clinical History	729	3.00
Ophthalmologic Disorders	703	10.00	Lab. Diagnosis of Inf. Dis.	965	7.50
JOURNAL ARTICLE COMPILATIONS			Math for Med Techs	964	7.00
Ambulance Service Journal Articles	517	10.00	Multilingual Guide for Medical Personnel	961	2.50
Blood Banking & Immunohemat. Jour. Art.	798	10.00	Neoplasms of the Gastrointestinal Tract	736	20.00
Emergency Room Journal Articles	795	8.00	Neurophysiology Study Guide	600	7.00
Hodgkin's Disease Journal Articles	515	12.00	Nursing & the Nephrology Patient	376	5.00
Hosp. & Inst. Eng. & Maintenance J. Art.	793	8.00	Outpatient Hemorrhoidectomy Lig. Tech.	752	12.50
Hosp. Electronic Data Process. J. Art.	791	8.00	Profiles in Surgery, Gynec. & Obstetrics	963	5.00
Hosp. Pharmacy Journal Articles	799	10.00	Radiological Physics Exam. Review	486	10.00
Hosp. Security & Safety Journal Articles	796	8.00	Skin, Heredity & Malignant Neoplasms	744	20.00
Human Cytomegalovirus Journal Articles	522	15.00	Testicular Tumors	743	20.00
Immunosuppressive Therapy Journal Art.	526	20.00	Tissue Adhesives in Surgery	756	24.00
			Understanding Hematology	977	8.00

Prices subject to change.

OTHER BOOKS AVAILABLE

ITEMS	Code	Unit Price	ITEMS	Code	Unit Price
MEDICAL EXAM REVIEW BOOKS			**STATE BOARD REVIEW BOOKS**		
Vol. 1 Comprehensive	101	$12.00	Med. State Brd. Rev. - Basic Sciences	411	$9.00
Vol. 2 Clinical Medicine	102	7.50	Med. State Brd. Rev. - Clinical Sciences	412	9.00
Vol. 2A Txtbk. Study Guide of Int. Med.	123	7.50	Cardiopulmonary Techn. Exam. Rev. - Vol. 1	473	7.50
Vol. 2B Txtbk. Study Guide of Int. Med.	130	7.50	Cytology Exam. Review Book - Vol. 1	454	7.50
Vol. 3 Basic Sciences	103	7.50	Dental Exam. Review Book - Vol. 1	431	7.50
Vol. 4 Obstetrics - Gynecology	104	7.50	Dental Exam. Review Book - Vol. 2	432	7.50
Vol. 4A Textbk. Study Guide of Gynecology	152	7.50	Dental Exam. Review Book - Vol. 3	433	7.50
Vol. 5 Surgery	105	7.50	Dental Hygiene Exam. Review - Vol. 1	461	7.50
Vol. 5A Textbk. Study Guide of Surgery	150	7.50	Emergency Med. Techn. Exam. Rev. - Vol. 1	465	7.50
Vol. 6 Public Health & Prev. Medicine	106	7.50	Emergency Med. Techn. Exam. Rev. - Vol. 2	466	7.50
Vol. 8 Psychiatry & Neurology	108	7.50	Immunology Exam. Review Book - Vol. 1	424	7.50
Vol.11 Pediatrics	111	7.50	Inhalation Therapy Exam. Review - Vol. 1	471	7.50
Vol.12 Anesthesiology	112	7.50	Inhalation Therapy Exam. Review - Vol. 2	344	7.50
Vol.13 Orthopaedics	113	10.00	Laboratory Asst. Exam. Rev. Bk. - Vol. 1	455	7.50
Vol.14 Urology	114	10.00	Medical Librarian Exam. Rev. Bk. - Vol. 1	495	7.50
Vol.15 Ophthalmology	115	10.00	Medical Record Library Science - Vol. 1	496	7.50
Vol.16 Otolaryngology	116	10.00	Medical Techn. Exam. Review - Vol. 1	451	7.50
Vol.17 Radiology	117	10.00	Medical Techn. Exam. Review - Vol. 2	452	7.50
Vol.18 Thoracic Surgery	118	10.00	Occupational Therapy Exam. Rev. - Vol. 1	475	7.50
Vol.19 Neurological Surgery	119	15.00	Optometry Exam. Review	469	10.00
Vol.20 Physical Medicine	128	10.00	Pharmacy Exam. Review Book - Vol. 1	421	7.50
Vol.21 Dermatology	127	10.00	Physical Therapy Exam. Review - Vol. 1	481	7.50
Vol.22 Gastroenterology	141	10.00	Physical Therapy Exam. Review - Vol. 2	482	7.50
Vol.23 Child Psychiatry	126	10.00	X-Ray Technology Exam. Rev. - Vol. 1	441	7.50
Vol.24 Pulmonary Diseases	143	10.00	X-Ray Technology Exam. Rev. - Vol. 2	442	7.50
Vol.25 Nuclear Medicine	133	10.00	X-Ray Technology Exam. Rev. - Vol. 3	443	7.50
Vol.26 Allergy	132	10.00	**NURSING EXAM REVIEW BOOKS**		
Vol.27 Plastic Surgery	129	10.00	Vol. 1 Medical-Surgical Nursing	501	4.50
Vol.28 Cardiovascular Diseases	138	10.00	Vol. 2 Psychiatric-Mental Health Nursing	502	4.50
Vol.29 Oncology	146	10.00	Vol. 3 Maternal-Child Health Nursing	503	4.50
ECFMG Exam Review - Part One	120	7.50	Vol. 4 Basic Sciences	504	4.50
ECFMG Exam Review - Part Two	121	7.50	Vol. 5 Anatomy and Physiology	505	4.50
BASIC SCIENCE REVIEW BOOKS			Vol. 6 Pharmacology	506	4.50
Anatomy Review	201	7.00	Vol. 7 Microbiology	507	4.50
Biochemistry Review	202	7.00	Vol. 8 Nutrition & Diet Therapy	508	4.50
Digestive System Basic Sciences	215	7.00	Vol. 9 Community Health	509	4.50
Heart & Vascular Systems Basic Sciences	212	7.00	Vol.10 History & Law of Nursing	510	4.50
Microbiology Review	203	7.00	Vol.11 Fundamentals of Nursing	511	4.50
Nervous System Basic Sciences	210	7.00	Practical Nursing Examination Rev. - Vol. 1	711	4.50
Pathology Review	204	7.00	**CASE STUDY BOOKS**		
Pharmacology Review	205	7.00	Allergy Case Studies	027	10.00
Physiology Review	206	7.00	Cardiology Case Studies	001	10.00
Respiratory System Basic Sciences	213	7.00	Chest Diseases Case Studies	012	10.00
Urinary System B.Sci.	214	7.00	Child Psychiatry Case Studies	029	10.00
Anatomy Textbook Study Guide	124	7.00	Cutaneous Medicine Case Studies	014	7.50
Histology Textbook Study Guide	151	7.00	ECG Case Studies	003	7.50
Medical Physiology Textbk. Study Guide	155	7.00	Endocrinology Case Studies	008	10.00
SPECIALTY BOARD REVIEW BOOKS			Gastroenterology Case Studies	004	10.00
Dermatology Specialty Board Review	311	10.00	Hematology Case Studies	020	10.00
Family Practice Specialty Board Review	309	10.00	Infectious Diseases Case Studies	011	7.50
Internal Medicine Specialty Board Review	303	10.00	Neurology Case Studies	006	10.00
Neurology Specialty Board Review	306	10.00	Orthopedic Surgery Case Studies	030	10.00
Obstetrics-Gynecology Spec. Bd. Review	304	10.00	Otolaryngology Case Studies	021	10.00
Pathology Specialty Board Review	305	10.00	Pediatric Hematology Case Studies	018	10.00
Pediatrics Specialty Board Review	301	10.00	Pediatric Oculo-Neural Dis. Case Studies	023	10.00
Psychiatry Specialty Board Review	312	10.00	Respiratory Care Case Studies	019	7.50
Surgery Specialty Board Review	302	10.00	Urology Case Studies	017	10.00
The Otolaryngology Boards	313	10.00	**MEDICAL OUTLINE SERIES**		
The Psychiatry Boards	307	8.00	Cancer Chemotherapy	631	10.00
			Child Psychiatry	613	10.00

Prices subject to change.